CONTENTS

ABOUT THE AUTHOR

Steven E. Barkan is Chairperson and Professor of Sociology at the University of Maine, where he has taught Introduction to Sociology and many other courses since 1979. His teaching and research interests include criminology, research methods, sociology of law, and social movements. Among his professional activities, he has served on the Board of Directors of the Society for the Study of Social Problems (SSSP), as chair of SSSP's Law and Society Division, as an Advisory Editor for SSSP's journal, *Social Problems*, and as a member of the Advisory Board for the American Sociological Association's Honors Program. His previous books include *Criminology: A Sociological Understanding* (2nd ed.), *Collective Violence* (with Lynne Snowden), and *Protesters on Trial: Criminal justice in the Southern Civil Rights and Vietnam Antiwar Movements*. He has also written numerous journal articles on such topics as household crowding and crime rates, racial prejudice and death penalty attitudes, feminist activism, social movement participation, and legal control of the Southern civil rights movement. These articles have appeared in the *American Sociological Review*, *Journal of Crime and Justice*, *Journal of Research in Crime and Delinquency*, *Race and Society*, *Social Forces*, *Social Problems*, *Sociological Forum*, and *Sociological Inquiry*. Professor Barkan welcomes comments from students and faculty about this workbook. You may reach him by e-mail at BARKAN@MAINE.EDU or by regular mail at Department of Sociology, 5728 Fernald Hall, University of Maine, Orono, ME 04469-5728.

PREFACE AND ACKNOWLEDGMENTS

Welcome to this innovative introduction to sociological discovery! The successful first edition of this book paved new ground in having students learn sociology by *doing* sociology with software that was simple to use but sophisticated in its approach. Students explored hundreds of social relationships as they discovered sociology. This second edition builds on the first by updating all the data sets and including new, more theoretically grounded exercises that build more directly upon previously learned skills and knowledge. As before, the exercises are geared to the chapter structure of the leading introduction to sociology textbooks.

As with the first edition, this workbook comes with four data sets: (1) GSS, the 2000 General Social Survey, administered to a large random sample of the U.S. population; (2) STATES, a compilation of data on each state derived from the 2000 U.S. Census and other sources; (3) NATIONS, a compilation of international survey and other data on most of the world's nations and population; and (4) HISTORY, trends in U.S. public opinion as measured by the General Social Survey since the early 1970s. With these data sets and the use of MicroCase's ExplorIt software, the workbook provides students a wonderful tool for learning the substance of sociology by doing their own sociological analysis. In every exercise, they will first see examples of data analysis on the standard topics central to the discipline of sociology, and they will then have the chance to do their own analysis using the four data sets. This active engagement with the sociological enterprise will help them learn about society in ways unimaginable little more than ten or fifteen years ago.

My interest in writing the first edition of this workbook stemmed from my first year in college in 1969, when I began learning sociology by analyzing numerical data on keypunch cards fed through a card sorter. This active engagement, as crude as it was by today's standards, with the process and logic of data analysis triggered my fascination with sociology. When MicroCase came along in the late 1980s, I eagerly began using it with my own students. The software has advanced considerably since that time and now represents a unique combination of simplicity and sophistication in social science data analysis software. Students have a world of sociological discovery awaiting just a few easy mouse clicks.

I am once again delighted to acknowledge the efforts of several people and organizations without whom this workbook would not have been possible. Norman Miller, my first sociology professor, introduced me to the fascination of empirical research in sociology. Writing this edition and its predecessor has been one way of repaying the immense debt I owe him. The MicroCase staff now working for the Wadsworth Publishing Company, and in particular Julie Aguilar, Jodi Gleason, and Matt Bahr, provided tremendous assistance in the preparation of this edition, as did Assistant Editor Analie

Barnett of Wadsworth. I also continue to owe David Smetters of the original MicroCase Corporation heartfelt thanks for first helping me to implement MicroCase's wonderful vision for introducing today's students to social science knowledge and data analysis.

I also would like to thank the sources of the data files accompanying this workbook. MicroCase's data archive, now under the auspices of Wadsworth, provided all four of these files, and the many hours involved in compiling them saved me even more hours of labor. Special thanks go to Tom W. Smith of the National Opinion Research Center for his continued direction and administration of the General Social Survey. Much of the international data in the book is based on the World Values Survey, for which I thank Ronald Inglehart at the Institute for Social Research, University of Michigan. Professor Inglehart's book, *Modernization and Postmodernization: Cultural, Economic and Political Change in 43 Societies* (Princeton University Press, 1997), inspired several of the examples in the workbook. Thanks also go to the Roper Center for its many years of distribution of the General Social Survey and to the Inter-university Consortium for Political and Social Research (ICPSR). The data archives maintained by ICPSR, the Roper Center, and Wadsworth/MicroCase provide an invaluable service to social science researchers and their students.

My next set of acknowledgments goes to the several instructors who reviewed the original workbook in preparation for this revision. Their comments were critical but fair and greatly improved the final product. Needless to say, any remaining errors are my responsibility. The reviewers are: Don Ecklund, Lincoln Land Community College; Christine Monnier, College of DuPage; Daniel V. A. Olson, Indiana University South Bend; Alden E. Roberts, Texas Tech University; Josephine A. Ruggiero, Providence College; Margaret Walsh, Keene State College.

My final acknowledgment goes to my family, Barb, Dave, and Joey. I dedicate this new workbook to them for their loving support over the years. As my two sons proceed through young adulthood, I hope their generous spirits and love of learning will bring them much success and help them meet the inevitable challenges that lie ahead.

GETTING STARTED

INTRODUCTION

Welcome to ExplorIt! With the easy-to-use software accompanying this workbook, you will have the opportunity to learn about sociology by exploring dozens of sociological issues with data from around the United States and around the world.

Each exercise in this workbook deals with a standard topic in typical sociology courses. The preliminary section of each exercise uses data from the workbook's software to illustrate key issues and social relationships related to that topic. You can easily create all the graphics in this part of the exercise by following the ExplorIt Guides you'll be seeing. Doing so will take just a few clicks of your computer mouse and will help you become more familiar with ExplorIt. The ExplorIt Guides are described in more detail below.

Each exercise also has a worksheet section where you'll do your own data analysis. This section usually contains about a dozen questions that will either follow up on examples from the preliminary section or have you explore some new issues. You'll use the workbook's software to answer these questions.

SYSTEM REQUIREMENTS

- Windows 95 (or higher)

- 8 MB RAM

- CD-ROM drive

- 15 MB of hard drive space (if you want to install it)

To run the software on a Macintosh, you will need emulation software or hardware installed. For more information about emulation software or hardware, check with your local Macintosh retailer or try the Web site http://machardware.about.com/cs/pcemulation/.

NETWORK VERSIONS OF STUDENT EXPLORIT

A network version of Student ExplorIt is available at no charge to instructors who adopt this book for their course. It's worth noting that Student ExplorIt can be run directly from the CD-ROM on virtually any computer network—regardless of whether a network version of Student ExplorIt has been installed.

INSTALLING STUDENT EXPLORIT

If you will be running Student ExplorIt directly from the CD-ROM—or if you will be using a version of Student ExplorIt that is installed on a network—skip to the section "Starting Student ExplorIt."

To install Student ExplorIt to a hard drive, you will need the CD-ROM that is packaged inside the back cover of this book. Then follow these steps in order:

To install Student ExplorIt to a hard drive, you will need the CD-ROM that is packaged inside the back cover of this book. Then follow these steps in order:

1. Start your computer and wait until the Windows desktop is showing on your screen.

2. Insert the CD-ROM disc into the CD-ROM drive of your computer.

3. On most computers the CD-ROM will automatically start and a welcome menu will appear. If the CD-ROM doesn't automatically start, do the following:

   Click [Start] from the Windows desktop, click [Run], type **D:\SETUP**, and click [OK]. (If your CD-ROM drive is not the D drive, replace the letter D with the proper drive letter.) To install Student ExplorIt to your hard drive, select the second option on the list: "Install Student ExplorIt to your hard drive."

4. During the installation, you will be presented with several screens, as described below. In most cases you will be required to make a selection or entry and then click [Next] to continue.

   The first screen that appears is the **License Name** screen. (If this software has been previously installed or used, it already contains the licensing information. In such a case, a screen confirming your name will appear instead.) Here you are asked to type your name. It is important to type your name correctly, since it cannot be changed after this point. Your name will appear on all printouts, so make sure you spell it completely and correctly! Then click [Next] to continue.

   A **Welcome** screen now appears. This provides some introductory information and suggests that you shut down any other programs that may be running. Click [Next] to continue.

   You are next presented with a **Software License Agreement**. Read this screen and click [Yes] if you accept the terms of the software license.

   The next screen has you **Choose the Destination** for the program files. You are strongly advised to use the destination directory that is shown on the screen. Click [Next] to continue.

5. The Student ExplorIt program will now be installed. At the end of the installation, you will be asked if you would like a shortcut icon placed on the Windows desktop. We recommend that you select [Yes]. You are now informed that the installation of Student ExplorIt is finished. Click the [Finish] button and you will be returned to the opening Welcome screen. To exit completely, click the option "Exit Welcome Screen."

STARTING STUDENT EXPLORIT

There are three ways to run Student ExplorIt: (1) directly from the CD-ROM, (2) from a hard drive installation, or (3) from a network installation. Each method is described below.

Starting Student ExplorIt from the CD-ROM

Unlike most Windows programs, it is possible to run Student ExplorIt directly from the CD-ROM. To do so, follow these steps:

1. Insert the CD-ROM disc into the CD-ROM drive.

2. On most computers the CD-ROM will automatically start and a Welcome menu will appear. (Note: If the CD-ROM does **not** automatically start after it is inserted, click [Start] from the Windows desktop, click [Run], type **D:\SETUP**, and click [OK]. If your CD-ROM drive is not the D drive, replace the letter D with the proper drive letter.)

3. Select the first option from the Welcome menu: **Run Student ExplorIt from the CD-ROM**. Within a few seconds, Student ExplorIt will appear on your screen. Type in your name where indicated to enter the program.

Starting Student ExplorIt from a Hard Drive Installation

If Student ExplorIt is installed to the hard drive of your computer (see earlier section "Installing Student ExplorIt"), it is **not** necessary to insert the CD-ROM. Instead, locate the Student ExplorIt "shortcut" icon on the Windows desktop, which looks something like this:

To start Student ExplorIt, position your mouse pointer over the shortcut icon and double-click (that is, click it twice in rapid succession). If you did not permit the shortcut icon to be placed on the desktop during the install process (or if the icon was accidentally deleted), you can alternatively follow these directions to start the software:

Click [Start] from the Windows desktop.

Click [Programs].

Click [MicroCase].

Click [Student ExplorIt - DS].

After a few seconds, Student ExplorIt will appear on your screen.

Starting Student ExplorIt from a Network

If the network version of Student ExplorIt has been installed to a computer network, you need to double-click the Student ExplorIt icon that appears on the Windows desktop to start the program. Type in your name where indicated to enter the program. (Note: Your instructor may provide additional information that is unique to your computer network.)

MAIN MENU OF STUDENT EXPLORIT

Student ExplorIt is extremely easy to use. All you do is point and click your way through the program. That is, use your mouse arrow to point at the selection you want, then click the left button on the mouse.

The main menu is the starting point for everything you will do in Student ExplorIt. Look at how it works. Notice that not all options on the menu are always available. You will know which options are available at any given time by looking at the colors of the options. For example, when you first start the software, only the OPEN FILE option is immediately available. As you can see, the colors for this option are brighter than those for the other tasks shown on the screen. Also, when you move your mouse pointer over this option, it is highlighted.

EXPLORIT GUIDES

Throughout this workbook, "ExplorIt Guides" provide the basic information needed to carry out each task. Here is an example:

> ➤ *Data File:* **STATES**
> ➤ *Task:* **Mapping**
> ➤ *Variable 1:* **56) MURDER**
> ➤ *View:* **Map**

Each line of the ExplorIt Guide is actually an instruction. Let's follow the simple steps to carry out this task.

Step 1: Select a Data File

Before you can do anything in Student ExplorIt, you need to open a data file. To open a data file, click the OPEN FILE task. A list of data files will appear in a window (e.g., GSS, HISTORY, NATIONS). If you click on a file name *once*, a description of the highlighted file is shown in the window next to this list. In the ExplorIt Guide shown above, the ➤ symbol to the left of the Data File step indicates that you should open the STATES data file. To do so, click STATES and then click the [Open] button (or just double-click STATES). The next window that appears (labeled File Settings) provides additional information about the data file, including a file description, the number of cases in the file, and the number of variables, among other things. To continue, click the [OK] button. You are now returned to the main menu of Student ExplorIt. (You won't need to repeat this step until you want to open a different data file.) Notice that you can always see which data file is currently open by looking at the file name shown on the top line of the screen.

Step 2: Select a Task

Once you open a data file, the next step is to select a program task. Six analysis tasks are offered in this version of Student ExplorIt. Not all tasks are available for each data file, because some tasks are appropriate only for certain kinds of data. Mapping, for example, is a task that applies only to ecological data, and thus cannot be used with survey data files.

In the ExplorIt Guide we are following, the ➤ symbol on the second line indicates that the MAPPING task should be selected, so click the MAPPING option with your left mouse button.

Step 3: Select a Variable

After a task is selected, you will be shown a list of the variables in the open data file. Notice that the first variable is highlighted and a description of that variable is shown in the Variable Description window at the lower right. You can move this highlight through the list of variables by using the up and down cursor keys (as well as the <Page Up> and <Page Down> keys). You can also click once on a variable name to move the highlight and update the variable description. Go ahead—move the high-light to a few other variables and read their descriptions.

If the variable you want to select is not showing in the variable window, click on the scroll bars located on the right side of the variable list window to move through the list. See the following figure.

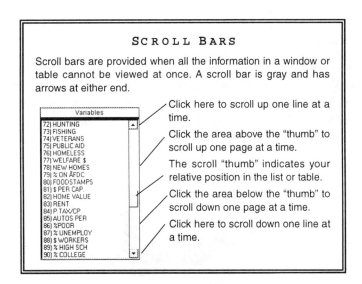

By the way, you will find an appendix at the back of this workbook that contains a list of the variable names for key data files provided in this package.

Each task requires the selection of one or more variables, and the ExplorIt Guides indicate which vari-ables you should select. The ExplorIt Guide example here indicates that you should select 56) MURDER as Variable 1. On the screen, there is a box labeled Variable 1. Inside this box, there is a vertical cursor that indicates that this box is currently an active option. When you select a variable, it will be placed in this box. Before selecting a variable, be sure that the cursor is in the appropriate box. If it is not, place the cursor inside the appropriate box by clicking the box with your mouse. This is important because in some tasks the ExplorIt Guide will require more than one variable to be selected, and you want to be sure that you put each selected variable in the right place.

To select a variable, use any one of the methods shown below. (Note: If the name of a previously selected variable is in the box, use the <Delete> or <Backspace> key to remove it—or click the [Clear All] button.)

- Type the **number** of the variable and press <Enter>.

- Type the **name** of the variable and press <Enter>. Or you can type just enough of the name to distinguish it from other variables in the data—MUR would be sufficient for this example.

- Double-click the desired variable in the variable list window. This selection will then appear in the variable selection box. (If the name of a previously selected variable is in the box, the newly selected variable will replace it.)

- Highlight the desired variable in the variable list, then click the arrow that appears to the left of the variable selection box. The variable you selected will now appear in the box. (If the name of a previously selected variable is in the box, the newly selected variable will replace it.)

Once you have selected your variable (or variables), click the [OK] button to continue to the final results screen.

Step 4: Select a View

The next screen that appears shows the final results of your analysis. In most cases, the screen that first appears matches the "view" indicated in the ExplorIt Guide. In this example, you are instructed to look at the Map view—that's what is currently showing on the screen. In some instances, however, you may need to make an additional selection to produce the desired screen.

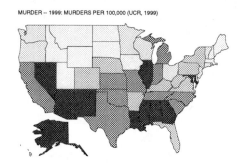

MURDER -- 1999: MURDERS PER 100,000 (UCR, 1999)

(OPTIONAL) Step 5: Select an Additional Display

Some ExplorIt Guides will indicate that an additional "Display" should be selected. In that case, simply click on the option indicated for that additional display. For example, this ExplorIt Guide may have included an additional line that required you to select the Legend display.

Step 6: Continuing to the Next ExplorIt Guide

Some instructions in the ExplorIt Guide may be the same for at least two examples in a row. For instance, after you display the map for population in the example above, the following ExplorIt Guide may be given:

> Data File: **STATES**
> Task: **Mapping**
> ➤ Variable 1: **54) V.CRIME**
> ➤ View: **Map**

Notice that the first two lines in the ExplorIt Guide do not have the ➤ symbol in front of the items. That's because you already have the data file STATES open and you have already selected the MAPPING task. With the results of your first analysis showing on the screen, there is no need to return to the main menu to complete this next analysis. Instead, all you need to do is select V.CRIME as your new variable. Click the [↺]] button located in the top left corner of your screen and the variable selection screen for the MAPPING task appears again. Replace the variable with 54) V.CRIME and click [OK].

To repeat: You need to do only those items in the ExplorIt Guide that have the ➤ symbol in front of them. If you start from the top of the ExplorIt Guide, you're simply wasting your time.

If the ExplorIt Guide instructs you to select an entirely new task or data file, you will need to return to the main menu. To return to the main menu, simply click the [Menu] button at the top left corner of the screen. At this point, select the new data file and/or task that is indicated in the ExplorIt Guide.

That's all there is to the basic operation of Student ExplorIt. Just follow the instructions given in the ExplorIt Guide and point and click your way through the program.

ONLINE HELP

Student ExplorIt offers extensive online help. You can obtain task-specific help by pressing <F1> at any point in the program. For example, if you are performing a scatterplot analysis, you can press <F1> to see the help for the SCATTERPLOT task.

If you prefer to browse through a list of the available help topics, select **Help** from the pull-down menu at the top of the screen and select the **Help Topics** option. At this point, you will be provided a list of topic areas. Each topic is represented by a closed-book icon. To see what information is available in a given topic area, double-click on a book to "open" it. (For this version of the software, use only the "Student ExplorIt" section of help; do not use the "Student MicroCase" section.) When you double-click on a book graphic, a list of help topics is shown. A help topic is represented by a graphic with a piece of paper with a question mark on it. Double-click on a help topic to view it.

If you have questions about Student ExplorIt, try the online help described above. If you are not very familiar with software or computers, you may want to ask a classmate or your instructor for assistance.

EXITING FROM STUDENT EXPLORIT

If you are continuing to the next section of this workbook, it is *not* necessary to exit from Student ExplorIt quite yet. But when you are finished using the program, it is very important that you properly exit the software—do not just walk away from the computer or remove your CD-ROM. To exit Student ExplorIt, return to the main menu and select the [Exit Program] button that appears on the screen.

Important: If you inserted your CD-ROM before starting Student ExplorIt, remember to remove it before leaving the computer.

THE SOCIOLOGICAL PERSPECTIVE

Tasks: Mapping
Data Files: STATES

We are all individuals, but we are also social beings influenced by our social environments. We grow up in a society and within a family in that society. We grow up as girls or boys, and, sooner than our parents would like, become women or men. We are members of a racial or ethnic group, or, depending on our parents' ancestry, more than one such group. We have low incomes, medium incomes, or high incomes. Many of us belong to a religious faith. We also grow up in different parts of the country and in urban, suburban, or rural areas in the country. All of these aspects of our social environment comprise our social background, and they all influence how we turn out: how we think, how we behave, and our chances for success or failure in life.

This is the fundamental truism of what is often called the *sociological perspective*: our social backgrounds influence our attitudes, behavior, and life chances. While no two people, even identical twins, are the same, neither are they completely different. If they've both grown up in the United States, they are automatically more similar to each other than if one had grown up in the U.S. and the other in Nigeria or Japan. If they are the same sex—both female or both male—they have more things in common than if one were female and the other male. All of this means that if we know enough about an individual's background, we can predict her or his attitudes, behavior, and eventual outcomes in life with surprising accuracy. We won't always be right, but we will be right more often than we're wrong.

This workbook's major goal is to illustrate the sociological perspective with real-life data drawn from the United States and around the world. You will learn sociology by doing sociology. You'll see again and again how social backgrounds influence behavior, attitudes, and life chances. You'll also see some surprising examples of expected influences not happening. Although much of our exploration will take place in the United States, we'll also be looking at a wide range of other nations, which differ in many ways that have important consequences for the attitudes, behavior, and life chances of their populations. Such a global perspective is increasingly important in today's world, not the least because it helps us understand our own society that much better.

Welcome, then, to our use of Student ExplorIt to discover society. The computer program itself is easy and even fun to use, and the hundreds of variables contained in the data sets accompanying this workbook provide you a vast storage of intriguing data to aid in your discovery.

This first exercise introduces you to the sociological perspective by using maps of the United States to illustrate regional differences in behavior and possible explanations for these differences. Later exercises will follow a different format. In them, we will first examine the exercise's subject matter with international data, then continue with data on the states of the United States, and end with data from a large, national survey of U.S. residents conducted in 2000. These three data sets will complement each other as you discover society.

SUICIDE AND THE SOCIOLOGICAL PERSPECTIVE

Suicide is one of the most tragic behaviors we can think of and also one of the hardest to explain. When we hear that someone committed suicide, a natural question to ask is "Why did she kill herself?" (or "Why did he kill himself?"). When we try to answer this question, we usually speculate that the suicide victim had been depressed over some unfortunate circumstance in his or her life. A marriage or relationship was ending, a job or school work was going poorly, the person's health was failing, and the like. All of these answers focus on the immediate, individual circumstances of the person's life, as well they should. What, after all, is a more individualistic act than suicide? What is more individualistic than deciding to take your own life?

A sociologist, however, approaches suicide as a social act, not just as an individual one. A sociologist asks, "Why is it that some *kinds* of individuals are more likely than other kinds to commit suicide?" To put it another way, "Why do some groups of individuals have higher *rates* of suicide than other groups?" The idea is that an individual from a group—or, more precisely, from a social background—with a higher rate of suicide is more likely to commit suicide than one from a group or social background with a low rate of suicide.

This central insight into the sociological perspective comes from the great French sociologist Emile Durkheim, one of the founders of sociology. Writing a century ago, Durkheim said that explanations focusing on individual unhappiness are insufficient to explain differences in group suicide rates. Instead, he said, there must be something about the group itself that explains its high or low rate. Suicide, then, does not just arise from forces inside the individual; it also arises from forces external to the individual. These forces are the properties of the group to which the individual belongs or the social background from which the individual comes.

Durkheim's focus on the influence of social forces on individual behavior and attitudes greatly influenced the development of the sociological perspective. Let's look at the United States today to see if some of the suicide patterns that Durkheim would have predicted still hold true.

We'll begin by using your STATES data set. Like the other data sets with this workbook, the STATES set contains many variables. In sociological jargon, a **variable** is anything that varies. To be more precise, it is any characteristic, feature, or dimension that takes on a different value among the things (or **units of analysis**) being studied. In the STATES data set, the unit of analysis is the state. We have 50 states in this data set, or 50 cases. States differ in many ways, and the STATES data set contains information on many of the ways, or variables, in which states differ. These variables include the size of each state's population, the percentage of each state's population that is poor or younger than age 18, and the circulation of *Field & Stream* magazine per 100,000 population.

As these examples indicate, many of the variables in the STATES data set are either percentages or rates. Both such figures take into account the fact that some states have more people than others. For example, to indicate how many poor people there are in California, a state of about 34 million people, and to compare that with the number of poor people in Kansas, a state with fewer than 3 million people, tells us little about which state is poorer, because the state's population sizes are so different. Sociologists and other social scientists thus use percentages or rates to compare geographic units such as states.

To calculate a *percent*, in case you have forgotten your early math lessons, divide the number of people who are poor (to stick with our poverty example) by the number of people living in a state and multiply the result by 100. Thus, to use some unrealistically small numbers to keep the math easier, if we have 25 poor people living in a state containing 200 people, we divide 25 by 200 to get .125 and then multiply this result by 100 to get 12.5 percent. This is equivalent to 12.5 poor people for every 100 poor people in the state, and we say that 12.5 percent of the state's population is poor.

A rate gives the same kind of information as a percent but commonly uses 1,000 or 100,000 (if we're still looking at states) instead of 100. We use these larger multipliers when calculating a percent would yield percent less than 1. Thus, if the number of murders is 250 in a state of 2,300,000 people, dividing 250 by 2,300,000 gives us .0001086. If we multiplied this figure by only 100, then we would say that .01086 percent of the state's population is murdered every year. Because murder is a rare event and thus yields percent much smaller than 1, we instead multiply the .0001086 figure by 100,000 to give us a murder rate of 10.86 per 100,000 population. (Note that a poverty rate of 12.5 per 100 population, or 12.5 percent, is the same as a rate of 12,500 per 100,000 population.)

Back to business. One of the variables in the STATES data set is the number of suicides per 100,000 population for each state. Because this variable uses a rate and not just the number of suicides, we can rank the states according to their rates. The higher the suicide rate in a state, the more common suicide is in that state, and the more likely it is that an individual in that state will commit suicide.

I'll now show the ExplorIt Guide that lists the steps to obtain the map for each state's suicide rate. If you don't understand the Guide, reread the "Getting Started" section at the front of this book. When you see ExplorIt Guides in this and all other exercises in the book, you're strongly advised to actually do the steps yourself at a computer. This procedure will help you learn how to use Student ExplorIt and become more familiar with the program. It will also help prepare you to complete the worksheets at the end of each exercise.

To repeat, our first ExplorIt Guide obtains a map depicting each state's suicide rate. Here we go!

➤ *Data File:* **STATES**
➤ *Task:* **Mapping**
➤ *Variable 1:* **80) SUICIDE**
➤ *View:* **Map**

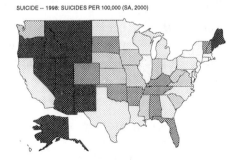

SUICIDE -- 1998: SUICIDES PER 100,000 (SA, 2000)

In all the maps we obtain in this workbook, the darker the color of a state (or, when we use the NATIONS data set, the darker the color of a nation), the "more" of a variable it has. Thus, in the map before you, the darker the color, the higher the state's suicide rate. When we look at this map, notice that the "darkest" states—those with the highest suicide rates—tend to be in the West.

The colors indicate which states have higher and lower suicide rates, but do not indicate the actual rate for each state. To determine this rate, you can click the [Legend] button at the top of the screen to see what range of rates each color stands for. This still doesn't yield the more precise information we need, so after we obtain a map, we will often obtain a list of each state's actual rate, percentage, or value. The ExplorIt Guide to obtain the ranking is:

Data File: **STATES**
Task: **Mapping**
Variable 1: **80) SUICIDE**
➤ View: **List: Rank**

RANK	CASE NAME	VALUE
1	Nevada	22.7
2	Alaska	21.0
3	Wyoming	18.1
4	Montana	17.9
5	Arizona	17.2
6	New Mexico	17.1
7	Oregon	16.6
8	Idaho	16.4
9	Utah	16.0
10	Maine	15.8

Inspect the rankings we've obtained. Not surprisingly, western states appear at the top of the list. Nevada leads the nation in suicide, with a rate of 22.7 per 100,000 people, and Alaska is just behind with a rate of 21.0. Wyoming, Montana, Arizona and New Mexico fill in the rest of the top six states for the rate of suicide. The lowest suicide rates, at the bottom of the list, are in New Jersey, with a rate of 7.2 suicides per 100,000 people, and New York, with a rate of 7.5. Connecticut ranks just above these two states.

Now that we know the West has the highest suicide rates, we need to try to explain this regional difference. We are trying to answer the question "Why does the West have the highest regional suicide rate?" If Durkheim is right, it's not enough to say that people in the West are more depressed than those in other regions, and we'd be hard put to prove that anyway. Sociologically speaking, there must be something else about life out West that increases the chances of suicide. Note that we're not saying that if you live in the West, you're probably going to commit suicide. Far from it! Even in Nevada, the state with the highest rate, "only" 22.7 of every 100,000 people commit suicide. Although that's the highest rate, you're far, far more likely not to commit suicide than to commit it if you live in Nevada.

Durkheim thought that suicide rates are higher among groups and in places with less social integration. One indicator of low social integration is what might be called "geographic mobility." The more people move around, the less social integration they have. Let's map a measure of the percent of each state's population not born in that state.

Data File: **STATES**
Task: **Mapping**
➤ Variable 1: **81) MOBILITY**
➤ View: **Map**

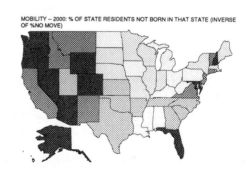

MOBILITY -- 2000: % OF STATE RESIDENTS NOT BORN IN THAT STATE (INVERSE OF %NO MOVE)

The darker the color, the greater the percent of each state's population who were not born in the state where they currently reside. The map is not identical to the previous one for suicide, but it is very similar. Generally the "darker" states—those with the highest geographic mobility and, presumably, the lowest social integration—are also in the West.

Yet another indicator of social integration is population density, or the number of people per square mile. Low population density would indicate low social integration, because it makes sense to think that the fewer people per square mile, the fewer connections they have with each other. Let's map a "lack of population density" variable such that the darker colors indicate the fewest people per square mile and the lighter colors the most people per square mile.

Data File: **STATES**	
Task: **Mapping**	
➤ *Variable 1:* **18) NOT DENSE**	
➤ *View:* **Map**	

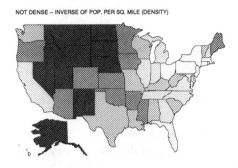

NOT DENSE -- INVERSE OF POP. PER SQ. MILE (DENSITY)

Remember, the darker the color, the lower the population density. Once again, we see a map similar to the suicide map. The West has the lowest population density and also the highest suicide rate.

Let's map one final variable, the percentage of each state's population who say they have no religion. All other things equal, if people don't go to religious services, they have fewer social networks and thus lower social integration. They may also be less likely to have a faith to turn to for comfort in times of personal trouble. All of this may help lead to higher suicide rates. If our hypothesis is correct, the percent with no religion should be highest in the West, and in our map the western states should have a darker color.

Data File: **STATES**	
Task: **Mapping**	
➤ *Variable 1:* **30) %NORELIG**	
➤ *View:* **Map**	

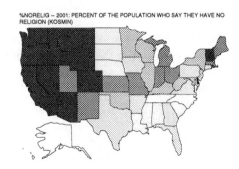

%NORELIG -- 2001: PERCENT OF THE POPULATION WHO SAY THEY HAVE NO RELIGION (KOSMIN)

This map again looks similar to the suicide one. Lack of religion is highest in the West, the region that also has the highest suicide rate.

Note that none of this proves Durkheim's theory of suicide. It is possible that suicidal, divorce-prone atheists move into the western states and then kill themselves. It is also possible that the western states differ in other ways than those mapped here that lead to higher suicide rates. Maybe people in all states are equally likely to try to kill themselves, but those in the West are better able to do so, perhaps because they have more guns. None of this is meant to make light of suicide, but rather to indicate the complexity of explaining social phenomena. Certainly the trends in our maps are consistent with predictions drawn from Durkheim's theory, but they do not prove the theory is correct.

The larger point is that an individual's chances of suicide might depend partly on where the individual lives, on various characteristics of that location, and on other aspects of the individual's social background. It might be true that only unhappy people commit suicide, but an unhappy person in the West is more apt to perform this act than an unhappy one elsewhere.

HEART DISEASE

According to the sociological perspective, our social backgrounds influence our life chances along with our behavior and attitudes. "Life chances" refer to our prospects for good or poor outcomes in life, and one of the most important life chances is the state of our health. Heart disease is one of our nation's most serious health problems. If our chances for suicide depend partly on where we live, might the same be true of heart disease? Let's examine the geographic distribution of the number of deaths from heart disease per 100,000 population.

Data File: **STATES**
Task: **Mapping**
➤ Variable 1: **36) HEART DTHS**
➤ View: **Map**

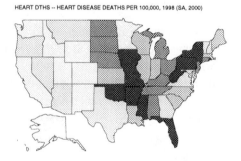

HEART DTHS -- HEART DISEASE DEATHS PER 100,000, 1998 (SA, 2000)

The rate of heart disease is generally higher east of the Mississippi River than west of it. Why should heart disease be higher in the eastern part of the nation? Does the eastern region have a higher percentage of people who are overweight and thus, from what medical experts tell us, are more susceptible to heart disease?

Data File: **STATES**
Task: **Mapping**
➤ Variable 1: **37) % FAT**
➤ View: **Map**

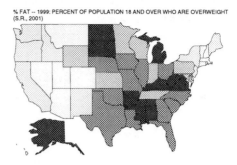

% FAT -- 1999: PERCENT OF POPULATION 18 AND OVER WHO ARE OVERWEIGHT (S.R., 2001)

People are indeed more likely to be overweight east of the Mississippi, perhaps accounting for the higher rate of heart disease found there. This still does not tell us why weight problems are more likely in the East, but it does indicate that where we live affects not only our chances of suicide but also our chances of contracting heart disease.

MAGAZINE CIRCULATION RATES

To take a far less serious topic, let's see if we can explain magazine circulation rates. Some states have higher rates than other states of readership of various magazines. As individuals, we all can decide whether or not to subscribe to a particular magazine, but perhaps our chances of doing so depend on where we live and other aspects of our social backgrounds. To illustrate this, examine the circulation rate for *Field & Stream* magazine. Do you ever read this magazine?

Data File: **STATES**
Task: **Mapping**
➤ Variable 1: **68) F&STREAM**
➤ View: **Map**

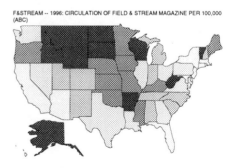

F&STREAM -- 1996: CIRCULATION OF FIELD & STREAM MAGAZINE PER 100,000 (ABC)

Field & Stream circulation is highest in the upper Midwest and mountain states and in northern New England.

Why should this magazine be so popular in these regions of the country? Does living in a rural area promote interest in *Field & Stream*? Look again at the "lack of population density" map presented earlier. Note how similar this map was to the *Field & Stream* map. Not surprisingly, *Field & Stream* magazine is most popular in the least densely populated, or more rural, states. Where you live is related not only to the grievous decision to commit suicide and to the risk of heart disease, but also to the less consequential outcome of which magazine you choose to read.

Let's look at one more magazine, the very controversial *Playboy*. Where do you think its circulation rates are highest?

Data File: **STATES**
Task: **Mapping**
➤ Variable 1: **67) PLAYBOY**
➤ View: **Map**

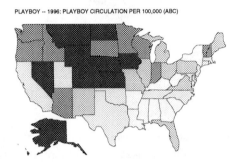

PLAYBOY -- 1996: PLAYBOY CIRCULATION PER 100,000 (ABC)

Playboy generally seems most popular west of the Mississippi, with the notable exception of Utah. What explains this pattern? If you look back at our earlier map for the percent of each state's population saying they have no religion, you'll see it's fairly similar to the *Playboy* map. Not surprisingly, *Playboy* seems to have its greatest appeal in the states whose populations are least religious.

We'll be dealing with more weighty topics in the rest of the workbook, but it should be clear by now that no matter how much we think we're independent individuals, our choice of magazines depends to some extent on where we live and on certain aspects of our social backgrounds such as our religious beliefs. If something as trivial as magazine choice and as serious as suicide and heart disease all illustrate the sociological perspective, will other aspects of our social lives be any exception? We'll find out in the exercises ahead.

WORKSHEET

NAME:

COURSE:

DATE:

EXERCISE

1

REVIEW QUESTIONS

Based on the first part of this exercise, answer True or False to the following items:

The sociological perspective emphasizes the influence of our individual personalities on our behavior.	T	F
Durkheim thought that external forces help explain suicide rates.	T	F
Suicide rates are higher in the West than in other regions of the United States.	T	F
Population density is higher in the West than in other regions of the United States.	T	F
Field & Stream magazine is most popular in the South.	T	F
The sociological perspective implies that people are totally determined by their social environments.	T	F

EXPLORIT QUESTIONS

You will need to use the ExplorIt software for the remainder of the questions. Make sure you have already gone through the "Getting Started" section that appears before the first exercise. If you have any difficulties using the software to obtain the appropriate informa- tion, or if you want to learn additional features of the MAPPING task, refer to the online help section [F1].

1. Let's examine murder rates in the United States. We will map the number of murders per 100,000 population.

> ➤ *Data File:* **STATES**
> ➤ *Task:* **Mapping**
> ➤ *Variable 1:* **56) MURDER**
> ➤ *View:* **Map**

   a. Which of the following regions has the highest murder rate? (Circle one.)

   Midwest

   New England

   South

   b. Which of the following regions has the lowest murder rate? (Circle one.)

   Midwest

   Far West

   South

2. Now let's look at the states' actual murder rates.

> *Data File:* **STATES**
> *Task:* **Mapping**
> *Variable 1:* **56) MURDER**
> ➤ *View:* **List: Rank**

  a. Which state has the highest murder rate?   _____

  b. What is this state's rate?   _____

  c. Which state has the lowest murder rate?   _____

  d. What is this state's rate?   _____

  e. What is the murder rate of the state in which your college or university is located?   _____

  f. Is this state's murder rate in the top half of the country or in the bottom half?   Top   Bottom

3. What factors might explain the regional differences in murder rates you have just seen? Do you think climate might matter? Let's find out.

> *Data File:* **STATES**
> *Task:* **Mapping**
> ➤ *Variable 1:* **4) WARM WINTR**
> ➤ *View:* **Map**

The darker the state, the warmer its January temperatures.

  a. Which of the following regions is the warmest? (Circle one.)   Midwest

        New England

        South

  b. Which of the following regions is the least warm? (Circle one.)   Midwest

        Far West

        South

  c. Does this map look similar to or very different from the murder map?   Similar   Very Different

  d. Do the warmer states tend to have higher murder rates?   Yes   No

e. What sociological reason might explain the answer you just gave?

4. Now we will examine another very important topic, poverty, in the United States. We will map the percent of each state's population that is below the poverty line.

> Data File: **STATES**
> Task: **Mapping**
> ➤ Variable 1: **45) %POOR**
> ➤ View: **Map**

a. Which region appears to be the poorest? (Circle one.)

Northeast

Midwest

South

b. Did this region also have a high murder rate?   Yes   No

5. While we're looking at poverty, let's determine which states are the poorest and which states are the least poor.

> Data File: **STATES**
> Task: **Mapping**
> Variable 1: **45) %POOR**
> ➤ View: **List: Rank**

a. List the three poorest states, and list the percent of each state's population that is poor.

STATE	% POOR
_____	_____
_____	_____
_____	_____

b. List the three states with the lowest poverty rates, starting with the one with the lowest rate, and also list the percent of each state's population that is poor.

STATE	% POOR
_____	_____
_____	_____
_____	_____

6. States also differ in their high school dropout rates. Will the states with the highest dropout rates also tend to be the poorest ones?

> Data File: **STATES**
> Task: **Mapping**
> ➤ Variable 1: **52) HS DROPOUT**
> ➤ View: **Map**

a. Which region has a higher high school dropout rate? (Circle one.)   Midwest   South

b. Did this region also have a high poverty rate?   Yes   No

7. Let's see which states have the highest dropout rates and which have the lowest ones.

> Data File: **STATES**
> Task: **Mapping**
> Variable 1: **52) HS DROPOUT**
> ➤ View: **List: Rank**

a. List the three states with the highest high school dropout rates, and list the percent of each state's population that dropped out of high school.

STATE	% DROPPED OUT
_____	_____
_____	_____
_____	_____

b. List the three states with the lowest dropout rates, starting with the one with the lowest rate, and also list the percent of each state's population that dropped out.

STATE   % DROPPED OUT

_____   _____

_____   _____

_____   _____

8. Use your answers to Questions 1–7 to answer the following questions:

a. The South generally has a higher murder rate than New England.   T   F

b. The state with the highest murder rate is New York.   T   F

c. Regions with high poverty rates also tend to have high murder rates.   T   F

d. Regions with high murder rates also tend to be relatively warm.   T   F

e. Minnesota has the lowest high school dropout rate.   T   F

f. Judging from the maps, poverty does not seem to be related to the
    high school dropout rate.   T   F

9. In the preliminary section of this exercise, you examined regional variations in *Field & Stream* magazine. Let's look at the subscription rate of yet another magazine, *Cosmopolitan*, which is designed to appeal to young, educated women. We should thus expect *Cosmopolitan* to be especially popular in states with high percentages of college-educated people. Your task here is to determine if there is, in fact, a similarity between states with high *Cosmopolitan* subscription rates and states with high percentages of college-educated people. These are the steps you should follow:

a. Create a map of 66) COSMO.

b. Obtain a ranked list of 66) COSMO. Print out this ranked distribution and turn it in with your assignment.

c. Next, create a map of 51) COLLEGE

d. Obtain a ranked list of 51) COLLEGE. Print out this ranked distribution and turn it in with your assignment.

e. As your final step, compare the maps for both variables and determine whether they look similar. In particular, determine whether the regions of the U.S. with high *Cosmopolitan* subscription rates are also those with high percentages of college-educated people, and whether the regions with low subscription rates are also those with low percentages of college-educated people. Next, compare the rankings you printed out for both variables. Determine whether the states that are high on one list tend to be high on the other list, and whether the states that are low on one list tend to be low on the other list. As a way of doing this, fill in the ranks (i.e., 1st, 2nd, 44th, 49th) for each of the following states:

	51) COLLEGE	66) COSMO
CONNECTICUT	_____	_____
MASSACHUSETTS	_____	_____
NEW JERSEY	_____	_____
ARKANSAS	_____	_____
MISSISSIPPI	_____	_____
WEST VIRGINIA	_____	_____

f. Which conclusion best describes the results of your investigation?

1. *Cosmopolitan* seems most popular in states with higher levels of education.
2. *Cosmopolitan* seems most popular in states with lower levels of education.
3. The level of education in the states does not seem to be related to *Cosmopolitan* subscription rates.

◆ EXERCISE 2 ◆

CULTURE AND SOCIETY

Tasks: Mapping, Scatterplot, Univariate
Data Files: NATIONS, STATES, GSS

Every society has its own culture, which is both material and nonmaterial. *Material culture* refers to the tangible objects that make up every society and includes such things as eating utensils, clothing, and artwork. Societies differ in the types of material culture they contain. *Nonmaterial culture* refers to the language, symbols, values, beliefs, and behaviors that characterize a society and that the society's members learn through socialization. If the culture we learn influences our attitudes, behaviors, and values, then culture is a key concept of the sociological perspective.

This exercise sketches some key cultural differences throughout the world and within the United States. As you develop your appreciation of culture's importance, think of the many ways in which you've been influenced by the culture in which you've grown up.

CULTURE AROUND THE WORLD

Let's start with various aspects of a society's material culture and see how nations differ in the types of objects found in them. Our first object will be one with which you're very familiar, the television.

> ➤ *Data File:* **NATIONS**
> ➤ *Task:* **Mapping**
> ➤ *Variable 1:* **44) TLVSN/CP**
> ➤ *View:* **Map**

TLVSN/CP – 1994: TELEVISIONS PER 10,000 POPULATION (UNSY, 1997)

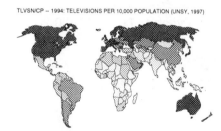

> To reproduce this graphic on the computer screen using ExplorIt, review the instructions in the "Getting Started" section. For this example, you would open the NATIONS data file, select the MAPPING task, and select 44) TLVSN/CP for variable 1. The first view shown is the Map view. (Remember, the ➤ symbol indicates which steps you need to perform if you are doing all examples as you follow along in the text. So, in the example that follows, you need only to select a new view—that is, you don't need to repeat the first three steps, because they were already done in this example.)

Remember, the darker the color, the more TVs per 10,000 population. Generally, North America and Europe have the most TVs, and Africa and Asia the least.

<table>
<tr><td>Data File:</td><td>NATIONS</td></tr>
<tr><td>Task:</td><td>Mapping</td></tr>
<tr><td>Variable 1:</td><td>44) TLVSN/CP</td></tr>
<tr><td>➤ View:</td><td>List: Rank</td></tr>
</table>

RANK	CASE NAME	VALUE
1	United States	8173.38
2	Malta	7391.30
3	Canada	6825.58
4	Japan	6800.49
5	Oman	6678.00
6	France	5889.26
7	Germany	5575.09
8	Denmark	5359.20
9	New Zealand	5193.31
10	Finland	5112.07

The United States leads the world, with about 8,173 TVs for every 10,000 people, while Malta (in southern Europe), Canada, Japan, and Oman (in the Middle East) rank two through five. At the other extreme are nations, most of them in Africa, with hardly any TVs. Comoros actually has none, while Eritrea has only 3 TVs for every 10,000 people.

Now let's examine nonmaterial culture. We'll start with an important behavior, attending religious services. The World Values Survey (WVS) data that are part of the NATIONS data set include the percentage of each nation's population who attend religious services at least monthly.

<table>
<tr><td>Data File:</td><td>NATIONS</td></tr>
<tr><td>Task:</td><td>Mapping</td></tr>
<tr><td>➤ Variable 1:</td><td>72) CH.ATTEND</td></tr>
<tr><td>➤ View:</td><td>Map</td></tr>
</table>

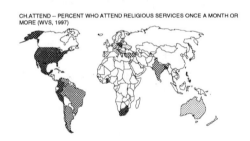

CH.ATTEND -- PERCENT WHO ATTEND RELIGIOUS SERVICES ONCE A MONTH OR MORE (WVS, 1997)

Religious attendance appears to be high in the United States and several other nations around the world, and relatively low in much of Europe.

<table>
<tr><td>Data File:</td><td>NATIONS</td></tr>
<tr><td>Task:</td><td>Mapping</td></tr>
<tr><td>Variable 1:</td><td>72) CH.ATTEND</td></tr>
<tr><td>➤ View:</td><td>List: Rank</td></tr>
</table>

RANK	CASE NAME	VALUE
1	Philippines	89.9
2	Ghana	82.4
3	Bangladesh	76.5
4	Poland	73.8
5	South Africa	71.2
6	Colombia	66.6
7	Mexico	65.2
8	Peru	64.1
9	United States	57.0
10	Dominican Republic	55.3

In the Philippines, 90 percent of the public attends religious services at least monthly for the highest rate in the data set. Ghana and Bangladesh rank next, at 82 and 77 percent respectively. Attendance in the United States is a bit lower at 57 percent. At the other end, only 9 percent of Estonia's population attends religious services at least monthly. Religion is clearly more a part of some nations' cultures than other nations' cultures.

Since we just viewed the map for religious attendance, let's look at the map for a variable that shows the percentage who say God is important in their lives.

> Data File: **NATIONS**
> Task: **Mapping**
> ➤ Variable 1: **73) GOD IMPORT**
> ➤ View: **Map**

GOD IMPORT -- PERCENT SAYING GOD IS IMPORTANT IN THEIR LIVES (WVS, 1997)

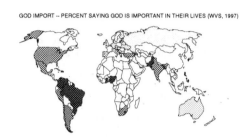

We would expect this map to look similar to the religious attendance one, and it does. Europe again ranks fairly low on this measure of religious belief, and several other nations report much higher percentages of belief in God's importance. Once again we see that religion plays a more important role in some nations' cultures than in other nations' cultures.

We can compare the maps for religious attendance and God's importance a bit more quickly with the following ExplorIt Guide.

> Data File: **NATIONS**
> Task: **Mapping**
> ➤ Variable 1: **72) CH.ATTEND**
> ➤ Variable 2: **73) GOD IMPORT**
> ➤ Views: **Map**

CH.ATTEND -- PERCENT WHO ATTEND RELIGIOUS SERVICES ONCE A MONTH OR MORE (WVS, 1997)

r = 0.840**

If you are continuing from the previous example, return to the variable selection screen for the MAPPING task. Select 72) CH.ATTEND for variable 1 and 73) GOD IMPORT for variable 2.

Now we again see the two maps we've already seen, but both at once.

It certainly makes sense to think that the countries where people are most likely to say God is important in their lives are also the countries where religious attendance is highest, which is what the two maps indicate. But how similar do two maps have to appear for us to say that they're similar? How different do they have to appear from each other for us to conclude that they're different? When we just look at the maps visually, it's sometimes difficult to determine precisely just how similar or different they are.

Fortunately, there is a simple method for determining how similar or different any two maps are. This method is called the **scatterplot** and was invented a century ago by English scholar Karl Pearson. You may have encountered a scatterplot when you learned algebra, but if you did not, or if you do not remember whether you did, don't let that worry you. It's easy to understand.

To develop a scatterplot, we first draw a horizontal line to represent the map of God's importance. Sweden ranked lowest on this variable, with only 26 percent of its population saying God is important in their lives. At the left end of our horizontal line we write 26 to indicate Sweden. At the right end we write the number 100 to represent Pakistan and Ghana, which had the highest percentage, 100, citing God's importance. Our horizontal line looks like this:

26   100

To represent the map of religious attendance, we now draw a vertical line up the left side of the horizontal line (to form an "L"). At the bottom of the new line we write 8 to represent the percent of Estonia's population that attends religious services at least monthly, since Estonia had the lowest percent. At the top we write 90 to represent Philippine's percent, which ranked highest.

We now have a figure that looks like this:

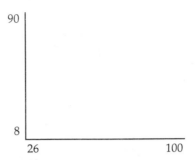

Now that we have a line with an appropriate scale to represent each map, we next look back at the rankings for each map to learn the value for each nation and then locate it on each line according to its score. Let's start with the Philippines. For the religious attendance variable, the Philippines had the highest percent, 90, so we can easily find its place on the vertical line. Make a small mark at 90 to locate the Philippines. For the God's importance variable, the Philippines ranked sixth at 97 percent, so we make a mark on the horizontal line just to the left of the 100 that is already there (for Pakistan and Ghana). Next we draw a line up from the mark for the Philippines on the horizontal line and draw another line to the right from the mark for the Philippines on the vertical line. Where these two lines intersect we draw a dot. This dot represents the combined map locations of the Philippines.

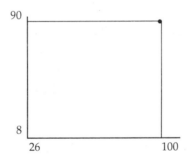

Now let's find the United States. Its value for the God's importance variable was 84, so we estimate where 84 is on the horizontal line and make a mark at that spot. The U.S. percentage for the religious attendance variable was 57, so we make a mark on the vertical line where 57 would be. We next draw a line up from the mark on the horizontal line and one out to the right from the mark on the vertical line. Where these lines meet is the combined map location for the United States. It's below and to the left of that for Nigeria.

When we have followed this procedure for each nation, we will have 40 dots, one for each nation, located within the space defined by the vertical and horizontal lines representing the two maps. We will have created a scatterplot. Fortunately, ExplorIt does all the work for you with just a few mouse clicks. The ExplorIt Guide that follows shows you how it's done.

Data File: **NATIONS**
➤ Task: **Scatterplot**
➤ Dependent Variable: **72) CH.ATTEND**
➤ Independent Variable: **73) GOD IMPORT**

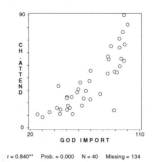

r = 0.840**   Prob. = 0.000   N = 40   Missing = 134

Notice that the scatterplot requires two variables.

Special Feature: When the scatterplot is showing, you may obtain the information on any dot by clicking on it. A little box will appear around the dot, and the values of 72) CH.ATTEND (or the x-axis variable) and of 73) GOD IMPORT (or the y-axis variable) will be shown.

Each of these dots represents a nation. We can see that the nations with the highest percentages on the God's importance variable (the ones to the right of the scatterplot) also tend to be the nations with the highest percentages on the religious attendance variable (the ones nearer to the top of the scatterplot).

Once Pearson created the scatterplot, his next step was to calculate what he called the **regression line**.

Data File: **NATIONS**
Task: **Scatterplot**
Dependent Variable: **72) CH.ATTEND**
Independent Variable: **73) GOD IMPORT**
➤ Display: **Reg. Line**

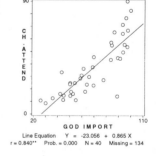

Line Equation   Y = -23.056 + 0.865 X
r = 0.840**   Prob. = 0.000   N = 40   Missing = 134

To show the regression line, select the [Reg. Line] option from the menu.

The regression line represents the line that best summarizes the trend of the dots; it's the line that comes closest to connecting all the dots. The closer the dots in any scatterplot are to the regression line, the more alike are the two maps represented in the scatterplot. The farther the dots are from the scatterplot and, thus, the less they resemble a straight line, the less the variables represented by the two maps have anything to do with each other.

To make it simpler to interpret the dots in a scatterplot, Pearson invented a statistic called the **correlation coefficient** and used the letter r as the symbol for this coefficient. The correlation coefficient varies from 0.0 to 1.0. When two maps are identical (as when you have two maps of the same variable), the correlation coefficient will be 1.0. When the maps are completely different from each other, the coefficient will be 0.0. The closer the correlation coefficient is to 1.0, then, the more alike the two maps.

Look at the lower left of the screen above and you will see r = 0.840. This indicates that the maps are very similar. Ignore the asterisks after this figure for now.

Correlation coefficients can be positive or negative. The one we've just seen is positive: when the God's importance percent is higher, the religious attendance percent is higher. As one variable goes up, so does the other. But negative correlations are also possible.

               Data File: **NATIONS**
                 Task: **Scatterplot**
➤ *Dependent Variable:* **8) FERTILITY**
➤ *Independent Variable:* **116) EDUCATION**
         ➤ *Display:* **Reg. Line**

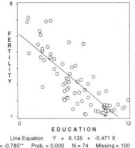

Our dependent variable is the average number of children born to a woman in her lifetime, and our independent variable is the average number of years of school a nation's adults have attended. Not surprisingly, the more educated a nation's population, the lower its fertility rate. Notice that the regression line now slopes downward from left to right, rather than upward, as in the previous example. This type of slope always indicates a negative correlation. Thus you'll also notice that a minus sign now precedes the correlation coefficient: r = –0.785.

Some correlations are above zero but are still too small for us to conclude that two variables are related. When this is the case, we treat the correlations as if they were a random accident and conclude that the correlation is in effect zero. The software automatically tells you whether the correlation was an accident or, instead, whether it was most likely due to a real relationship between the variables. If you look back at the correlation between God's importance and religious attendance, you'll see that two asterisks follow the value of r (r = 0.840**). Two asterisks means there is less than 1 chance in 100 that this correlation is a random accident. One asterisk means there is less than 1 chance in 20 that it's an accident. When no asterisks follow a correlation coefficient, the chances are too high that it could be a random accident, and we conclude there is no correlation at all. *Treat all correlations without asterisks as zero correlations.*

When r has at least one asterisk, it indicates a statistically significant relationship. In assessing the strength of this correlation, treat an r smaller than .3 (in absolute value) as a weak relationship, an r between .3 and .6 as a moderate relationship, and an r greater than .6 as a strong relationship.

Keep in mind, though, that correlation does not necessarily mean causation. Sometimes two variables can be correlated without one affecting the other. For example, in the months when more ice cream is sold, the crime rate is higher. Does that mean eating ice cream causes crime? Does it mean that fear of crime causes people to eat ice cream? Of course not! Obviously more ice cream is sold in summer months, and, for quite different reasons, crime is higher during the summer months. As another example, the more people listen to rock music, the worse their acne is. Does that mean that listening to rock music excites your pores and causes you to break out? Could it mean that having bad acne forces you to stay home and listen to rock to make you feel better? Not at all! Obviously younger people have worse acne, and younger people are more apt for different reasons to listen to rock music. A correlation between two variables might indicate a cause-and-effect relationship, but it doesn't have to.

A correlation between two variables also doesn't automatically tell us which variable is affecting which. This is called the "causal order" or "chicken and egg" question. In the religion example above, we saw a strong relationship between God's importance and religious attendance. But it's not clear which variable is causing which. Are people going to religious services more often because they believe God is more important in their lives, or do they believe God is more important in their lives because they're going to religious services more often? Or could both "causal directions" be valid?

CULTURE IN THE UNITED STATES

A few maps will indicate the regional distribution of aspects of the U.S. material culture. We'll start with the percent of occupied housing structures lacking a telephone.

> Data File: **STATES**
> Task: **Mapping**
> Variable 1: **29) NO PHONE**
> View: **Map**

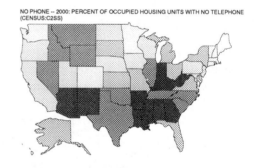

NO PHONE -- 2000: PERCENT OF OCCUPIED HOUSING UNITS WITH NO TELEPHONE (CENSUS:C2SS)

Notice that the STATES data file must be open.

The darker the color, the greater the percent of a state's households without a telephone. The South has the highest rate of households without a telephone. What accounts for regional variation in an essential object in the U.S. material culture?

Data File: **STATES**
Task: **Mapping**
Variable 1: **29) NO PHONE**
> View: **List: Rank**

RANK	CASE NAME	VALUE
1	New Mexico	6.19
2	Mississippi	5.92
3	Arkansas	4.84
4	West Virginia	4.72
5	Georgia	4.53
6	Alabama	4.38
7	Indiana	4.09
8	Louisiana	3.94
9	Kentucky	3.86
10	Arizona	3.84

Discovering Sociology

New Mexico has the highest rate of nonphone households, 6.19 percent. At the other end, only 0.8 percent of New Hampshire's households lack a telephone.

In some parts of the country, the pick-up truck is very much a part of the material culture of a local area or even that of an entire state. Let's see where the pick-ups are.

Data File: **STATES**
Task: **Mapping**
➤ Variable 1: **44) PICKUPS**
➤ View: **Map**

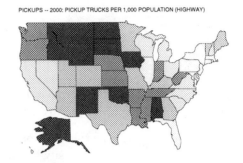

PICKUPS -- 2000: PICKUP TRUCKS PER 1,000 POPULATION (HIGHWAY)

There seem to be more pick-up trucks west of the Mississippi River, and especially in the northern part of that region of the country.

Data File: **STATES**
Task: **Mapping**
Variable 1: **44) PICKUPS**
➤ View: **List: Rank**

RANK	CASE NAME	VALUE
1	Wyoming	443
2	Montana	353
3	Alaska	291
4	North Dakota	289
5	Idaho	287
6	South Dakota	260
7	New Mexico	237
8	Oklahoma	233
9	Alabama	231
10	Iowa	229

Wyoming has 443 pick-ups per 1,000 people, but New York has only 30 per 1,000 people.

Earlier we asked what accounts for the regional variation in the lack of telephones in the United States. It makes sense to think that poorer states will be more likely to lack phones. Let's check this out with a scatterplot.

Data File: **STATES**
➤ Task: **Scatterplot**
➤ Dependent Variable: **29) NO PHONE**
➤ Independent Variable: **45) %POOR**
➤ Display: **Reg. Line**

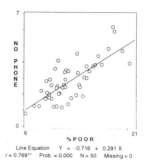

The poorer a state, the higher its rate of nontelephone households. The correlation coefficient, r, is a high 0.77**. (We have rounded r to two digits after the decimal point and will do that from now on.)

Now let's see whether we can explain the presence of pick-up trucks. We'll predict that pick-ups will be more common where population density, the number of people per square mile, is lowest.

Data File: **STATES**
Task: **Scatterplot**
➤ Dependent Variable: **44) PICKUPS**
➤ Independent Variable: **10) DENSITY**
➤ Display: **Reg. Line**

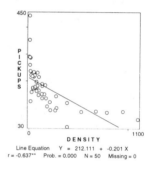

The lower the population density, the higher the pick-up rate (r = –0.64**).

We can also gauge the U.S. culture with survey questions given to a national, random sample of the U.S. adult population. The General Social Survey is one such survey and has been administered regularly since 1972; this workbook uses the 2000 version. A random sample of the U.S. population is just what it sounds like. As a sample, it's a much smaller subset of the entire population; the number of people in the 2000 GSS sample is 2,817. It's random because everyone in the country has the same chance of being in the sample. Although fewer than 3,000 end up in the sample, we all had the same chance of being included. Random samples save a lot of time and money, because we can gauge attitudes and behaviors of the entire adult population, some 200 million, with fewer than 3,000 respondents. Because the sample is random, we can generalize the results from the sample to the entire population. More on that in the next exercise.

One value often associated with the United States is the belief in the importance of democracy. The GSS asked its respondents how satisfied they are with the way democracy works in the United States.

> *Data File:* **GSS**
> *Task:* **Univariate**
> *Primary Variable:* **151) SAT DEMOC?**
> *View:* **Pie**

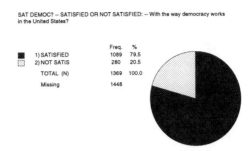

SAT DEMOC? -- SATISFIED OR NOT SATISFIED: -- With the way democracy works in the United States?

	Freq.	%
1) SATISFIED	1089	79.5
2) NOT SATIS	280	20.5
TOTAL (N)	1369	100.0
Missing	1448	

Notice that the GSS data file must be open.

Almost 80 percent of respondents are satisfied with the way democracy works, and about 20 percent are not satisfied. Clearly satisfaction with democracy is part of the American culture.

Another value in the U.S. culture is freedom, which has been valued ever since the colonial days. Americans like to make their own choices and don't like having people telling them what to do. The GSS asks respondents how important the following statement is to them: "Freedom is having the power to choose and do what I want in life."

> *Data File:* **GSS**
> *Task:* **Univariate**
> *Primary Variable:* **165) CHOICE**
> *View:* **Pie**

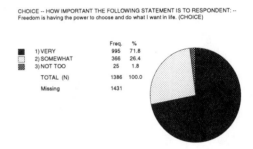

CHOICE -- HOW IMPORTANT THE FOLLOWING STATEMENT IS TO RESPONDENT: -- Freedom is having the power to choose and do what I want in life. (CHOICE)

	Freq.	%
1) VERY	995	71.8
2) SOMEWHAT	366	26.4
3) NOT TOO	25	1.8
TOTAL (N)	1386	100.0
Missing	1431	

Almost 72 percent of respondents say this statement is very important. The spirit of the colonial period still lives in America.

WORKSHEET

NAME: _____

COURSE: _____

DATE: _____

EXERCISE

2

REVIEW QUESTIONS

Based on the first part of this exercise, answer True or False to the following items:

Telephones and television are examples of nonmaterial culture.	T F
Religious attendance is higher in Europe than in North America.	T F
In the United States, households in the Midwest are especially likely not to have phones.	T F
In the United States, the poorer a state, the higher the rate of households without telephones.	T F
Judging from the GSS, less than half of Americans are satisfied with the way democracy works.	T F
A nonzero correlation always tells us that one variable is affecting another variable.	T F

EXPLORIT QUESTIONS

> **If you have any difficulties using the software to obtain the appropriate information, or if you want to learn additional features of the MAPPING or SCATTERPLOT tasks, refer to the online help section [F1].**

1. People of different nationalities living in the same society sometimes have different subcultures. Let's see which states have the highest percentages of people with Hispanic ancestry.

> ➤ *Data File:* **STATES**
> ➤ *Task:* **Mapping**
> ➤ *Variable 1:* **17) %HISPANIC**
> ➤ *View:* **Map**

   a. Which region of the country generally has the highest percentage of residents with Hispanic ancestry? (Circle one.)

Northeast

South

Midwest

West

Exercise 2: Culture and Society

b.   What might account for the regional patterns that you see?

2.   To continue our look at Hispanic ancestry, let's look at each state's ranking.

> Data File: **STATES**
> Task: **Mapping**
> Variable 1: **17) %HISPANIC**
> ➤ View: **List: Rank**

a.   Which state has the highest percent with Hispanic ancestry?   _____

b.   What is its percent?   _____%

c.   Which state has the lowest percent with Hispanic ancestry?   _____

d.   What is its percent?   _____%

3.   Now we'll look at the states in terms of their proportions of college graduates and their median family incomes.

> Data File: **STATES**
> Task: **Mapping**
> ➤ Variable 1: **51) COLLEGE**
> ➤ Variable 2: **46) MED.FAM. $**
> ➤ Views: **Map**

a.   These maps are (circle one):

Almost identical

Similar, but it is somewhat
difficult to determine how similar

Not very similar

Almost opposite

b.   Which region generally has the highest proportion of college
graduates? (Circle one.)

Northeast

South

Midwest

c. Which region has the lowest proportion overall? (Circle one.)

Northeast

South

Midwest

d. Which region generally has the highest median family
income? (Circle one.)

Northeast

South

Midwest

e. Which region has the lowest income overall? (Circle one.)

Northeast

South

Midwest

4. Now we'll compare these maps by using a scatterplot.

> Data File: **STATES**
> ➤ Task: **Scatterplot**
> ➤ Dependent Variable: **51) COLLEGE**
> ➤ Independent Variable: **46) MED.FAM. $**
> ➤ Display: **Reg. Line**

a. The states that are highest on 51) COLLEGE should appear
as dots at the (circle one):

Right of the scatterplot

Left of the scatterplot

Top of the scatterplot

Bottom of the scatterplot

b. The dots appearing at the bottom of the scatterplot represent
the states that have the (circle one):

Highest median incomes

Lowest median incomes

Highest percent of college graduates

Lowest percent of college graduates

c. What is the value of r for this scatterplot?

r = _____

d. Is r statistically significant?

Yes   No

e. The scatterplot indicates that states with higher family
incomes (circle one):   Have higher percentage of college graduates

                                                 Have lower percentage of college graduates

                                                 Have about the same percentage of college
graduates as states with low family incomes

f. Can you conclude from the scatterplot that income affects the chances of graduating from college?
1. Yes, because the states with higher incomes also have higher proportions of college graduates
2. No, because it is possible that states have higher incomes because they have higher proportions of college graduates

5. Recall that we examined the distribution of telephone ownership in the United States. Let's now look at phone ownership around the world, using a measure of the number of telephone lines per 1,000 population.

       ➤ *Data File:* **NATIONS**
          ➤ *Task:* **Mapping**
       ➤ *Variable 1:* **41) TELPH/CP**
          ➤ *Display:* **Map**

a. Overall, which of the following areas has the highest rate of telephone ownership? (Circle one.)   North America

                                                       South America

                                                             Africa

b. Which has the lowest rate of phone ownership? (Circle one.)   North America

                                                       South America

                                                             Africa

6. To continue our exploration, let's see how the various nations rank on this variable.

           Data File: **NATIONS**
              Task: **Mapping**
       Variable 1: **41) TELPH/CP**
          ➤ *View:* **List: Rank**

a. Which country has the highest rate of telephone ownership?   _____

b. What is its rate?   _____

c. Which nation has the lowest rate of telephone ownership?   _____

d. What is this nation's rate? _____

e. Does the United States rank in the top ten? Yes   No

7. Recall that the poorest states in the United States had the lowest rates of phone ownership. Let's hypothesize that we'll find a similar link between poverty and phone ownership around the world. Our independent variable will be the annual domestic product per capita, a rough measure of a nation's wealth or poverty.

> Data File: **NATIONS**
> ➤ Task: **Scatterplot**
> ➤ Dependent Variable: **41) TELPH/CP**
> ➤ Independent Variable: **30) GDP/CAP**
> ➤ Display: **Reg. Line**

Examine this scatterplot and note its correlation coefficient. In the space below, write a paragraph stating the hypothesis we're testing, reporting the results of your analysis, and drawing a conclusion about whether the data support this hypothesis.

8. The work ethic is a value that is often said to be an important part of the American nonmaterial culture. The GSS asks whether people get ahead by their own hard work or, instead, by luck or help from others. If the work ethic is that important, a majority of the public should say hard work is how people get ahead. Let's see if this is true.

> ➤ Data File: **GSS**
> ➤ Task: **Univariate**
> ➤ Primary Variable: **88) GET AHEAD?**
> ➤ View: **Pie**

a. What percent say people get ahead through hard work? _____%

b. What percent say people get ahead by being lucky? _____%

c. Do these results support the belief that the work ethic is an important value in American society? Yes   No

9. Related to the work ethic is the belief that American society allows people to improve their lot in life if they work hard enough. The GSS asks whether respondents agree that "people like me have a good chance of improving our standard of living." Let's see whether a majority of the public agrees with this statement.

Data File: **GSS**
Task: **Univariate**
➤ Primary Variable: **132) GOOD LIFE**
➤ View: **Pie**

a. What percent of the sample agrees that they have a good chance of
   improving their standard of living?   _____%

b. Did you think this percent would be higher than it was, lower than it was, or about what you
   found? Explain your answer.

c. Using your results for the GOOD LIFE and GET AHEAD? variables, how would you characterize
   the American culture?

d. In your opinion, how realistic are the views expressed through these two variables? Is life pretty
   much like what the majority of the public thinks?

◆ EXERCISE 3 ◆

SOCIALIZATION

Tasks: Mapping, Scatterplot, Univariate, Cross-tabulation, Auto-Analyzer
Data Files: NATIONS, GSS

If people are to become social beings and not just individuals, they must first learn what their society expects of them. They must learn their society's norms, values, and symbols, and other aspects of their society's culture. Socialization is the process by which people learn all of these things and more. Although socialization begins in infancy and is perhaps most important during the many years before adulthood, it continues throughout our mature years as well. A society without socialization is impossible.

This exercise looks at some aspects of socialization around the world and within the United States. Our international focus will be on values that people in different nations think children should learn. We'll see that the importance people place on these values differs among nations and is related to ways in which nations differ. We'll also look at these views about socialization with the GSS sample and use other variables to illustrate how our parents influence our thinking and behavior.

INTERNATIONAL DIFFERENCES IN SOCIALIZATION

Several countries in the NATIONS data set were asked whether it's important that children learn four different values: (1) good manners, (2) independence, (3) obedience, and (4) thriftiness. Let's map each of these.

➤ *Data File:* **NATIONS**
➤ *Task:* **Mapping**
➤ *Variable 1:* **112) KID MANNER**
➤ *View:* **Map**

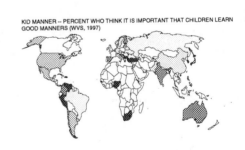

KID MANNER -- PERCENT WHO THINK IT IS IMPORTANT THAT CHILDREN LEARN GOOD MANNERS (WVS, 1997)

The importance placed on children's learning good manners certainly varies around the world, and no clear geographic pattern really emerges.

Data File: **NATIONS**

Task: **Mapping**

➤ Variable 1: **113) KID INDEPN**

➤ View: **Map**

Again, not the clearest pattern, but Japan and several nations in Europe seem most likely to think that children need to develop independence.

Data File: **NATIONS**

Task: **Mapping**

➤ Variable 1: **114) KID OBEY**

➤ View: **Map**

The highest percentages here are found in South America, Mexico, Africa, and India. The only exception to this pattern is the high ranking for the United Kingdom.

The final socialization variable is the percent who say it's very important that a child learn to be thrifty.

Data File: **NATIONS**

Task: **Mapping**

➤ Variable 1: **115) KID THRIFT**

➤ View: **Map**

There is a clear pattern here where the highest percentages are found in Eastern Europe and Asia. Mexico also ranks high on this question.

What explains these international differences in the perceived importance of different values for children's socialization? As just one example, do you think that when people are more educated, they should be more likely to value independence? If so, the level of a nation's education should be related to the importance its citizens place on independence. Let's find out.

Data File: **NATIONS**
➤ Task: **Scatterplot**
➤ Dependent Variable: **113) KID INDEPN**
➤ Independent Variable: **116) EDUCATION**
➤ Display: **Reg. Line**

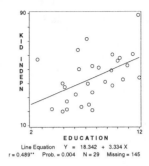

The more educated a nation's population, the more likely they are to say that it's very important for children to learn independence (r = .49**). Our hypothesis is supported.

SOCIALIZATION IN THE UNITED STATES

The way parents "should" socialize their children is a matter of continuing controversy. A few decades ago, when mothers of young children became more likely to work outside the home, many observers applauded this move, while others thought the children would be worse off without a full-time parent at home. What does the U.S. public think today about this issue? The GSS asked respondents whether they agree or disagree that "a preschool child is likely to suffer if his or her mother works." Do you agree with this statement? Why or why not? Compare your response to the GSS results.

➤ Data File: **GSS**
➤ Task: **Univariate**
➤ Primary Variable: **104) PRESCH.WRK**
➤ View: **Pie**

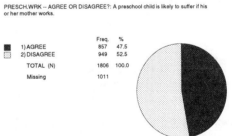

The public is almost evenly divided. Slightly less than half think that preschool children do suffer if their mothers work, while slightly more than half think they do not suffer.

Do you think gender will make a difference in whether people think preschool children fare worse if their mothers work? Does it make sense to think that women will be less apt than men to feel this way, perhaps because it's women whose careers are being challenged by this view? Let's find out. We'll first obtain the distribution of our "preschool" variable just for women. To do so, we'll use ExplorIt's subset variable capability.

Data File:	**GSS**	
Task:	**Univariate**	
Primary Variable:	**104) PRESCH.WRK**	
➤ Subset Variable:	**19) GENDER**	
➤ Subset Category:	**Include: 1) Female**	
➤ View:	**Pie**	

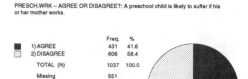

PRESCH.WRK -- AGREE OR DISAGREE?: A preschool child is likely to suffer if his or her mother works.

		Freq.	%
■	1) AGREE	431	41.6
▨	2) DISAGREE	606	58.4
	TOTAL (N)	1037	100.0
	Missing	551	

[Subset]

The option for selecting a subset variable is located on the same screen you use to select other variables. For this example, select 19) GENDER as a subset variable. A window will appear that shows you the categories of the subset variable. Select 1) Female as your subset category and choose the [Include] option. Then click [OK] and continue as usual.

With this particular subset selected, the results will be limited to the females in the sample. The subset selection continues until you exit the task, delete all subset variables, or clear all variables.

When we look only at women, we see that 41.6 percent think preschool children suffer if their mothers work. Now let's look only at men.

Data File:	**GSS**	
Task:	**Univariate**	
Primary Variable:	**104) PRESCH.WRK**	
Subset Variable:	**19) GENDER**	
➤ Subset Category:	**Include: 2) Male**	
➤ View:	**Pie**	

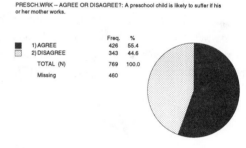

PRESCH.WRK -- AGREE OR DISAGREE?: A preschool child is likely to suffer if his or her mother works.

		Freq.	%
■	1) AGREE	426	55.4
▨	2) DISAGREE	343	44.6
	TOTAL (N)	769	100.0
	Missing	460	

[Subset]

The easiest way to change the subset category to Males (from Females) is to first delete the subset variable 19) GENDER. Then reselect 19) GENDER as the subset variable. Include 2) Male as your subset category. Then click [OK] and continue as usual.

When we look just at men, we find that 55.4 percent think preschool children suffer if their mothers work. If we subtract the 41.6 percent for women from the 55.4 percent for men, we see that there is a 13.8-percentage-point difference and that women are less likely than men to think that preschool children fare worse if their mothers work.

We found this gender difference using a two-step process that involves obtaining two univariate distributions. Your software includes a one-step procedure for obtaining the same result. This procedure is called a **_cross-tabulation_** and yields a table including the same data you saw in the two univariate distributions. Here's how it looks.

Discovering Sociology

Data File: **GSS**
➤ *Task:* **Cross-tabulation**
➤ *Row Variable:* **104) PRESCH.WRK**
➤ *Column Variable:* **19) GENDER**
➤ *View:* **Tables**
➤ *Display:* **Column %**

| | | GENDER | | |
		FEMALE	MALE	TOTAL
P R E S C H . W R K	AGREE	431	426	857
		41.6%	55.4%	47.5%
	DISAGREE	606	343	949
		58.4%	44.6%	52.5%
	Missing	551	460	1011
	TOTAL	1037	769	1806
		100.0%	100.0%	

To construct this table, return to the main menu and select the CROSS-TABULATION task. Then select 104) PRESCH.WRK as the row variable and 19) GENDER as the column variable. When the table is showing, select the [Column %] option.

The table indicates that 41.6 percent of women think preschool children suffer if their mothers work, compared to 55.4 percent of men. This matches the univariate results obtained above, as it should.

Remember that the GSS is a sample, or a subset, of the larger U.S. population. Does this 13.8-percent-age-point difference reflect an actual difference in the entire population, or is it something we got just by accident? If we flipped 1,000 coins, you know that we "should" get 500 heads, but you also know that someone could easily get 482 heads and someone else 510 heads. By the same token, a sample can sometimes differ from a population for random reasons. Because of this, we need to determine whether we can assume that any cross-tabulation's result does indeed reflect what's happening in the population, or whether it's just a random accident.

Sociologists and other social scientists use tests of statistical significance to make this determination. Differences observed in a random sample are said to be statistically significant when these differences are high enough that they are not likely to be a random accident, and thus reflect a real difference in the population from which the sample is drawn. In sociology, we assume that any observed difference in a sample is statistically significant when it occurs fewer than 5 times out of 100 by chance alone. When we say that a difference in a sample is statistically significant, we conclude that it really exists in the population from which the sample is drawn. If a difference would occur more than 5 times out of 100 by chance alone (a probability level of .05), we say that the difference is not statistically significant, and we conclude that it does not exist in the population from which the sample is drawn.

Some sociologists think a .05 probability level is too high and instead prefer a .01 level. That is, they assume that a difference is statistically significant only if it would occur less than 1 time out of 100 by chance alone.

To see what the level of significance is for the table we calculated above, look near the top where it reads "V = .137." The letter V stands for Cramer's V, which is a statistic indicating the strength of a relationship in cross-tabulations. Consider it the equivalent of Pearson's r, which you used in the previous exercise. If V is followed by one asterisk, the difference it represents is statistically significant at the .05 level of probability. That is, it would be expected to occur fewer than 5 times out of 100 by chance alone. If V is followed by two asterisks, the difference it represents is statistically significant at the .01 level of probability; it would be expected to occur less than 1 time out of 100 by chance alone. Whether V has one or two asterisks after it, we conclude the difference in the sample is statistically significant and thus exists in the entire U.S. population. If V is followed by *no* asterisks, the difference it represents is not statistically significant, and we conclude that there is no difference in the U.S. population.

Notice that the V of .137 in our table is followed by two asterisks. This means that the gender difference in the table would occur by chance alone less than 1 time out of 100. We can conclude with high confidence that the gender difference exists in the entire U.S. population.

Sometimes in large samples such as the GSS, a difference of only 2 or 3 percentage points can be statistically significant. This difference is small and, in fact, "nothing to write home about," even if it exists in the entire population. As this example indicates, a difference that is statistically significant might not be *substantively significant*. When you examine cross-tabulations in this book, always inspect the actual percentage differences to see how big a relationship really is. When you try to assess the strength of a relationship, also look at the size of Cramer's V. Any V greater than .30 should be considered a strong relationship, and any V between .10 and .30 should be considered a moderate relationship. A V under .10 should be considered a weak relationship. In the example above, V is .137, so we'd conclude that gender is moderately related to the view about preschool children and working mothers.

We can also get the table we just examined by using the AUTO-ANALYZER task. This task combines the univariate and cross-tabulation procedures you've already seen in these pages. It first shows you the distribution of a *primary variable* you select and then allows you to choose one of thirteen demographic variables—marital relationships, age, region, gender, race, political party, education, Southern residence, religious preference, family income, own (personal) income, denomination, and race/gender categories—to see what difference, if any, this demographic variable makes. It then gives you the appropriate cross-tabulation and actually tells you what is happening in the table. Let's see how it works.

Data File: **GSS**
➤ Task: **Auto-Analyzer**
➤ Variable: **104) PRESCH.WRK**
➤ View: **Univariate**

PRESCH.WRK -- AGREE OR DISAGREE?: A preschool child is likely to suffer if his or her mother works.

	%
AGREE	47.5%
DISAGREE	52.5%
Number of cases	1806

Among all respondents, 47.5% of the sample think a preschool child with suffer with a working mother.

To obtain these results, return to the main menu and select the AUTO-ANALYZER task. Then select 104) PRESCH.WRK as your variable and click [OK].

Here's the univariate distribution for 104) PRESCH.WRK. As you can see, the percentages are the same as those we found earlier when we first examined this variable using the UNIVARIATE task from the main menu. Now let's see what difference sex makes for these percentages.

Discovering Sociology

Data File: **GSS**
Task: **Auto-Analyzer**
Variable: **104) PRESCH.WRK**
➤ View: **Gender**

PRESCH.WRK -- AGREE OR DISAGREE?: A preschool child is likely to suffer if his or her mother works.

	FEMALE	MALE
AGREE	41.6%	55.4%
DISAGREE	58.4%	44.6%
Number of cases	1037	769

Among males, 55.4% think a preschool child will suffer with a working mother. Among females, this percentage was only 41.6%. The difference is statistically significant.

If continuing from the previous example, simply select [Gender] to see the results.

These results are identical to the ones we obtained using the CROSS-TABULATION task from the main menu, and the textual summary points to a gender difference in the quality considered most important for children to learn. As noted, the differences are statistically significant.

Let's use AUTO-ANALYZER to see whether education makes a difference and, if so, how it makes a difference. We saw with the NATIONS data set that the more educated a nation's populace, the more it favors independence in children and the less it favors obedience. This suggests that higher levels of education increase people's acceptance of less traditional beliefs. If that is true, then it makes sense to think that, within the United States, people with higher levels of education will be less likely than those with lower levels of education to think that preschool children suffer if their mothers work.

Data File: **GSS**
Task: **Auto-Analyzer**
Variable: **104) PRESCH.WRK**
➤ View: **Education**

PRESCH.WRK -- AGREE OR DISAGREE?: A preschool child is likely to suffer if his or her mother works.

	NOT H.S.	H.S. GRAD	SOME COLL.	COLL. GRAD.
AGREE	51.1%	50.0%	47.4%	42.5%
DISAGREE	48.9%	50.0%	52.6%	57.5%
Number of cases	319	508	521	449

Those who didn't graduate high school are most likely (51.1%) to think a preschool child will suffer with a working mother and those who graduated college (42.5%) are least likely. The difference is statistically significant.

If continuing from the previous example, simply select [Education] to see the results.

As AUTO-ANALYZER indicates, people with college educations are less likely (42.5 percent) than those without high school degrees (51.1 percent) to think that preschool children suffer. The result is statistically significant.

We can also use the GSS to illustrate how characteristics of our parents and aspects of our childhood influence the way we turn out. You might think that your religious preference, if you have one, is totally your individual choice, but have you ever thought that it's instead a product of your socialization by your parents? The GSS asks respondents to identify the religion in which they were raised. We can use this variable to see whether it predicts what religious preference respondents report as adults. We'll go back to the main menu and the CROSS-TABULATION task for this analysis.

Data File: **GSS**
➤ Task: **Cross-tabulation**
➤ Row Variable: **51) RELIGION**
➤ Column Variable: **55) RELIG. 1ST**
➤ View: **Tables**
➤ Display: **Column %**

RELIGION by RELIG. 1ST
Cramer's V: 0.861 **

		PROTESTANT	CATHOLIC	JEWISH	Missing	TOTAL
RELIGION	PROTESTANT	1346	106	3	66	1455
		96.4%	14.7%	5.4%		66.9%
	CATHOLIC	44	615	2	18	661
		3.2%	85.1%	3.6%		30.4%
	JEWISH	6	2	51	4	59
		0.4%	0.3%	91.1%		2.7%
	Missing	214	121	8	211	554
	TOTAL	1396	723	56	299	2175
		100.0%	100.0%	100.0%		

Does our current religious preference depend heavily on our parents' religious preference? Most certainly! Regardless of the religion in which we're raised, we're very likely to retain that religious preference when we're older (assuming we have one at all). Cramer's V is a very strong .86**. Although we may think that religious preference is an individual choice, the odds are very great that we keep the preference in which our parents raised us.

WORKSHEET

NAME:

COURSE:

DATE:

Workbook exercises and software are copyrighted. Copying is prohibited by law.

REVIEW QUESTIONS

Based on the first part of this exercise, answer True or False to the following items:

Socialization does not last beyond childhood.	T F
People in nations with higher levels of education are more likely than people in nations with less education to think that children should learn independence.	T F
In the GSS, women are more likely than men to believe preschool children suffer if their mothers work.	T F
Our parents' religious preference strongly affects our own religious preference.	T F
In the GSS, slightly more than half of the sample feels that preschool children suffer if their mothers work.	T F
GSS respondents who were raised as Protestants are very likely to continue to be Protestants.	T F

EXPLORIT QUESTIONS

> **If you have any difficulties using the software to obtain the appropriate information, or if you want to learn additional features of the MAPPING, SCATTERPLOT, UNIVARIATE, CROSS-TABULATION, or AUTO-ANALYZER tasks, refer to the online help section.**

1. Let's use AUTO-ANALYZER to see what other demographic variables, if any, are related to views about preschool children and working mothers.

> ➤ *Data File:* **GSS**
> ➤ *Task:* **Auto-Analyzer**
> ➤ *Variable:* **104) PRESCH.WRK**
> ➤ *View:* **Marital**

> **Open the GSS data file and select the AUTO-ANALYZER task. Select 104) PRESCH.WRK as your variable and click [OK]. When the table appears, select [Marital] to see the results.**

a. Widowed people are more likely than never-married people to think that preschool children suffer if their mothers work.   T   F

b. In this table, marital status is unrelated to this view about preschool children and working mothers.   T   F

c. Married people are more likely than people with any other marital status to think that preschool children suffer.   T   F

2. Now let's look at religion.

> Data File: **GSS**
> Task: **Auto-Analyzer**
> Variable: **104) PRESCH.WRK**
> ➤ View: **Religion**

a. Protestants are the most likely to think that preschool children suffer.   T   F

b. Catholics are the least likely to think that preschool children suffer.   T   F

c. This cross-tabulation does not illustrate the sociological perspective.   T   F

3. Should race make a difference?

> Data File: **GSS**
> Task: **Auto-Analyzer**
> Variable: **104) PRESCH.WRK**
> ➤ View: **Race**

a. African Americans are more likely than whites to think that preschool children suffer.   T   F

b. There is no racial difference in this view about preschool children and working mothers.   T   F

4. Drawing on the information in Questions 1–3, indicate which one of the following descriptions best summarizes the characteristics of people who are most likely to think that preschool children suffer if their mothers work.

    1. Married, Protestant, Black
    2. Widowed, Protestant, White
    3. Widowed, Jewish, White
    4. Divorced, Catholic, Black

5. Earlier in this exercise, we saw that respondents with higher education are less likely than those with lower education to think that preschool children suffer if their mothers work. Does it make sense to think that respondents with better-educated *parents* should also be less likely to think that preschool children suffer if their mothers work?

> *Data File:* **GSS**
> ➤ *Task:* **Cross-tabulation**
> ➤ *Row Variable:* **104) PRESCH.WRK**
> ➤ *Column Variable:* **163) PARS.DEGR.**
> ➤ *View:* **Tables**
> ➤ *Display:* **Column %**

To construct this table, return to the main menu and select the CROSS-TABULATION task, then select **104) PRESCH.WRK** as the row variable and **163) PARS.DEGR.** as the column variable. When the table is showing, select the **[Column %]** option. You may need to scroll the table to see all the columns.

a. What percent of people with college-educated parents (labeled BOTH COLL) think preschool children suffer? _____%

b. What percent of people with parents who did not graduate from high school (BOTH <HS) think those children suffer? _____%

c. Is V statistically significant?   Yes   No

d. Is our parents' education related to our own view about preschool children and working mothers?   Yes   No

6. How do the results in these worksheet exercises (Questions 1–5) illustrate the sociological perspective?

7. Let's see whether age is related to the belief that preschool children suffer if their mothers work.

> *Data File:* **GSS**
> *Task:* **Cross-tabulation**
> *Row Variable:* **104) PRESCH.WRK**
> ➤ *Column Variable:* **14) AGE 65+**
> ➤ *View:* **Tables**
> ➤ *Display:* **Column %**

a. What percent of people 65 and older say that preschool children suffer? _____%

b. What percent of people age 18–64? _____%

c. What is the value of V?   V = _____

d. Is V statistically significant? (Circle one.)   Yes   No

e. Which of the following is the most appropriate conclusion to draw from this cross-tabulation?

    1. Younger people are more likely than older people to believe that preschool children suffer if their mothers work.

    2. Older people are more likely than younger people to believe that preschool children suffer if their mothers work.

    3. Older people are less likely than younger people to believe that preschool children suffer if their mothers work.

8. Now let's use AUTO-ANALYZER to see whether the region of the country in which people live is related to their views on preschool children and working mothers.

>   Data File: **GSS**
> ➤   Task: **Auto-Analyzer**
> ➤   Variable: **104) PRESCH.WRK**
> ➤   View: **Region**

a. Which two regions are most likely to think that preschool children suffer
if their mothers work? (Circle one pair of regions.)

    Northeast and Midwest

    Northeast and South

    South and Northeast

    South and West

b. Reflecting on the answer you just gave about the results in AUTO-ANALYZER, how does this example illustrate the sociological perspective?

9. The GSS asks what is the ideal number of children for a family to have. Let's see whether the number of siblings respondents had when they were growing up is related to the number of children they see as ideal.

>   Data File: **GSS**
> ➤   Task: **Cross-tabulation**
> ➤   Row Variable: **91) IDEAL#KIDS**
> ➤   Column Variable: **11) # SIBS**
> ➤   View: **Tables**
> ➤   Display: **Column %**

a. What percent of respondents with four or more siblings think the ideal number
of children is three or more?   _____%

b. What percent of respondents with no siblings think the ideal number of children
is three or more?   _____%

c. Is V statistically significant?   Yes   No

d. The best conclusion to draw from this table is that

    1.   fewer siblings we have, the larger the family size we prefer.

    2.   more siblings we have, the larger the family size we prefer.

    3.   more siblings we have, the smaller the family size we prefer.

10. If we are indeed socialized by our parents' views and experiences, then whether our mother worked outside the home might influence our views on whether children in general are harmed if their mothers
work. Let's determine whether having had a working mother is related to our views on the issue of working mothers and their children. (Note: Examine each variable description before completing the analysis.)

            Data File: **GSS**
                 Task: **Cross-tabulation**
   ➤ *Row Variable:* **104) PRESCH.WRK**
➤ *Column Variable:* **24) MA WRK GRW**
           ➤ *View:* **Tables**
       ➤ *Display:* **Column %**

a. After examining the various percents in the table and V, briefly summarize the key results of this
table. What conclusion would you draw about whether and how having had a working mother
influences our own views on working mothers and their children?

11. Let's go back to the NATIONS data set and see how each nation ranks on the variable asking
whether it's very important for a child to learn obedience.

   ➤ *Data File:* **NATIONS**
       ➤ *Task:* **Mapping**
 ➤ *Variable 1:* **114) KID OBEY**
       ➤ *View:* **List: Rank**

a. Which nation ranks highest on this variable?   _____

b. What is the percent for this nation's population?   _____%

c. What is the percent for the United States?   _____%

d. Why do you think the United States does not rank higher than it does?

12. Do the more religious nations place more importance on children's learning obedience?

> Data File: **NATIONS**
> ➤ Task: **Scatterplot**
> ➤ Dependent Variable: **114) KID OBEY**
> ➤ Independent Variable: **73) GOD IMPORT**
> ➤ View: **Reg. Line**

a. What is the value of r?   r = _____

b. Is r statistically significant? (Circle one.)   Yes   No

c. Which statement below best describes the relationship depicted by this scatterplot?
    1. The more religious a nation's people, the more they believe that children should obey.
    2. The less religious a nation's people, the more they believe that children should obey.
    3. The more religious a nation's people, the less they believe that children should obey.

b. How does this example illustrate the importance of socialization?

GROUPS AND ORGANIZATIONS

Tasks: Mapping, Scatterplot, Univariate, Historical Trends, Cross-tabulation
Data Files: NATIONS, STATES, GSS, HISTORY

Four centuries ago, the English poet John Donne wrote, "No man is an island." To avoid sexist language, today one might write, "No one is an island." However it's put, Donne's observation illustrates keen sociological insight. Except for the occasional hermit, all of us have ties to other people. We have so many ties, in fact, that we're members of many different types of groups. Group life and group membership are essential to social life and are key to appreciating the sociological perspective.

Two basic types of groups exist: primary groups and secondary groups. ***Primary groups*** are small and involve strong social bonds. Common examples include your family, groups of best friends, juvenile gangs, and perhaps sports teams and fraternities and sororities. Ideally, primary groups provide us with a sense of belonging and of our identity, emotional support, and practical needs. ***Secondary groups*** tend to be larger than primary groups and involve weaker and much more impersonal social bonds. Common examples include any classes you've taken, including the one for which you're reading this book, and any business in which you might have worked. Secondary groups are not nearly as important as primary ones for our emotional needs, but they fulfill many of our practical needs and are certainly essential to any modern society. Many secondary groups are highly structured and are thus considered ***formal organizations***. Most formal organizations have clear lines of authority, explicit rules for behaving, and clearly defined roles for performing organizational tasks—in short, a division of labor.

This exercise explores some key aspects of group life and organizational membership around the world and within the United States. Its aim is to help you understand some of the key building blocks of modern society and thus to further your appreciation of the sociological perspective.

INTERNATIONAL EXAMPLES OF GROUP PERCEPTION AND MEMBERSHIP

Perhaps the most important primary group is the family, which we'll explore further in Exercise 10. However, it's also true that the family is more important to some people than to others within the same nation, and more important to the people of some nations than to those of other nations. Let's explore international differences in the importance attached to families. We'll use a measure from the NATIONS data set of the percent who say their families are "very important" in their lives.

➤ *Data File:* **NATIONS**
➤ *Task:* **Mapping**
➤ *Variable 1:* **111) FAMILY IMP**
➤ *View:* **Map**

FAMILY IMP -- PERCENT WHO SAY THE FAMILY IS VERY IMPORTANT IN THEIR LIVES (WVS, 1997)

No clear pattern really emerges here, but it does seem that the nations of Eastern Europe and Asia are least likely to feel their family is very important.

Data File: **NATIONS**
Task: **Mapping**
Variable 1: **111) FAMILY IMP**
➤ *View:* **List: Rank**

RANK	CASE NAME	VALUE
1	Macedonia	98.0
1	Venezuela	98.0
3	Turkey	97.5
4	Bosnia	97.2
5	Philippines	95.8
5	Nigeria	95.8
7	United States	95.2
8	Bangladesh	94.4
9	South Africa	93.9
10	Brazil	92.7

The residents of Macedonia and Venezuela, followed closely by those of Turkey, Bosnia, and the Philippines, are most likely to feel their family is very important; more than 90 percent in each nation report this view. At the other end, only 74 percent of the residents of Mexico, Germany, and Lithuania feel this way. Note that even with these differences, the world's people are still quite apt to feel their families are very important.

What accounts for this international difference in the importance placed on families? Perhaps families are considered more important in nations that are more religious.

Data File: **NATIONS**
➤ *Task:* **Scatterplot**
➤ *Dependent Variable:* **111) FAMILY IMP**
➤ *Independent Variable:* **73) GOD IMPORT**
➤ *Display:* **Reg. Line**

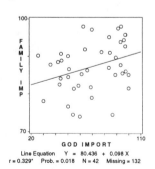

Line Equation   Y = 80.436 + 0.098 X
r = 0.329*   Prob. = 0.018   N = 42   Missing = 132

Our hypothesis is supported: the more religious a nation's citizens, the more they believe the family is very important in their lives (r = .33*). Why should the more religious nations feel this way about the family?

The NATIONS data set includes a variable on membership in one type of voluntary association, a sports or recreation group. The percent belonging to such a group varies by nation.

Data File: **NATIONS**
➤ *Task:* **Mapping**
➤ *Variable 1:* **109) DO SPORTS?**
➤ *View:* **Map**

DO SPORTS? -- PERCENT WHO SAY THEY BELONG TO A SPORTS OR RECREATION GROUP OR ORGANIZATION (WVS, 1997)

European nations have the highest rates of membership in sports or recreation groups.

What accounts for this international variation? Scholars say that education is a strong predictor of membership in many types of voluntary associations: the higher the education, the more associations joined. If this is true, then membership in sports or recreation groups should be higher in nations with more educated citizens.

Data File: **NATIONS**
➤ *Task:* **Scatterplot**
➤ *Dependent Variable:* **109) DO SPORTS?**
➤ *Independent Variable:* **116) EDUCATION**
➤ *Display:* **Reg. Line**

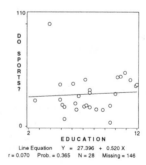

Line Equation   Y = 27.396 + 0.520 X
r = 0.070   Prob. = 0.365   N = 28   Missing = 146

At the international level, this assertion is not true. Membership in sports or recreation groups is not linked to education (r = .07).

ORGANIZATIONAL MEMBERSHIP IN THE UNITED STATES

One of the best-known types of voluntary associations is the Boy Scouts. One rough measure of membership in the Boy Scouts is the circulation of the magazine *Boys' Life*. Let's examine its circulation per 1,000 population.

> *Data File:* **STATES**
> *Task:* **Mapping**
> *Variable 1:* **53) BOYS' LIFE**
> *View:* **Map**

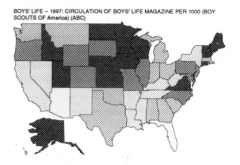

BOYS' LIFE -- 1997: CIRCULATION OF BOYS' LIFE MAGAZINE PER 1000 (BOY SCOUTS of America) (ABC)

This is an interesting geographic pattern. The circulation of *Boys' Life* magazine—and presumably interest and membership in the Boy Scouts—is clearly higher in the northern half of the country. What would you think accounts for this geographic pattern?

Data File: **STATES**
Task: **Mapping**
Variable 1: **53) BOYS' LIFE**
> *View:* **List: Rank**

RANK	CASE NAME	VALUE
1	Alaska	9
2	Kansas	8
2	North Dakota	8
2	Delaware	8
2	New Hampshire	8
2	Nebraska	8
2	Minnesota	8
2	Utah	8
9	Vermont	7
9	Iowa	7

Alaska leads the nation in *Boys' Life* circulation, and Arizona and Florida are tied with the lowest circulation.

This map gives us an idea of interest and membership in one kind of organization, but survey data are more often used to understand voluntary association and other secondary group membership in the United States. Accordingly, let's turn to the GSS, which contains several variables that will interest us.

A labor union certainly fits the definition of a secondary group. Let's see how many U.S. residents belong to a labor union.

> *Data File:* **GSS**
> *Task:* **Univariate**
> *Primary Variable:* **86) UNIONIZED?**
> *View:* **Pie**

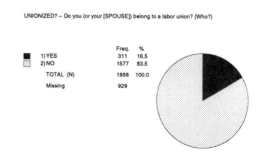

UNIONIZED? -- Do you (or your [SPOUSE]) belong to a labor union? (Who?)

	Freq.	%
■ 1) YES	311	16.5
▨ 2) NO	1577	83.5
TOTAL (N)	1888	100.0
Missing	929	

Just under 17 percent of Americans belong to a labor union.

We can use the HISTORICAL TRENDS task to see whether union membership has changed in the last 30 years.

> *Data File:* **HISTORY**
> > *Task:* **Historical Trends**
> *Variable:* **2) UNIONIZED**

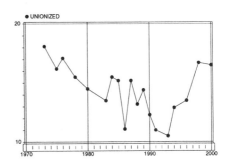

Percent of respondents belonging to a labor union

Open the HISTORY data file and select the HISTORICAL TRENDS task. Select 2) UNIONIZED as your trend variable.

Union membership has declined by a few percentage points since the early 1970s. Although it looks as though there is a lot of variation during the past 25 years, it is important to always look at the range displayed on the historical trends graph. Here you will see that union membership reached a high of about 18 percent in 1973 and a low of about 10.5% in 1992. This scale will always adjust depending on the range of the variables being used, so it is always important to check what the true range of the display is before interpreting the line.

Whether or not we belong to labor unions, we all have opinions about them. The GSS asks how much confidence respondents have in organized labor.

> *Data File:* **GSS**
> > *Task:* **Univariate**
> *Primary Variable:* **72) LABOR?**
> > *View:* **Pie**

LABOR? -- CONFIDENCE? Organized labor.

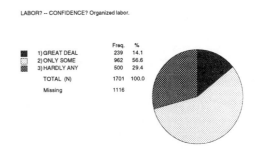

	Freq.	%
1) GREAT DEAL	239	14.1
2) ONLY SOME	962	56.6
3) HARDLY ANY	500	29.4
TOTAL (N)	1701	100.0
Missing	1116	

About 14 percent of Americans say they have a "great deal" of confidence in organized labor, 57 percent "only some" confidence, and 29 percent "hardly any" confidence.

Has the percentage expressing a great deal of confidence declined in the past 30 years along with union membership?

➤ *Data File:* **HISTORY**
  ➤ *Task:* **Historical Trends**
  ➤ *Trend 1:* **3) LABOR?**

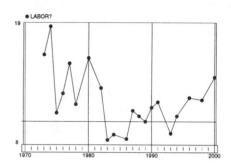

Percent expressing "great deal" of confidence in organized labor

Again, open the HISTORY data file and select the HISTORICAL TRENDS task. Select 3) LABOR? as your trend variable.

Confidence in organized labor declined until the mid-1980s and then rose somewhat, but is still not back to its early-1970s level.

The GSS also asks whether respondents are members of any group "whose main aim is to preserve or protect the environment." Are you a member of such a group? How many Americans do you think belong to one?

➤ *Data File:* **GSS**
  ➤ *Task:* **Univariate**
➤ *Primary Variable:* **166) GREEN GRP?**
  ➤ *View:* **Pie**

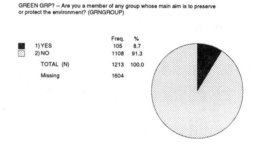

Although environmental groups are certainly often in the news, less than 9 percent of the GSS sample belongs to one of these groups.

The GSS also asks whether respondents have given money to an environmental group. Perhaps more people have donated money to environmental groups than are actually members.

Data File: **GSS**

Task: **Univariate**

➤ Primary Variable: **167) GREEN MONY**

➤ View: **Pie**

GREEN MONY -- In the last five years, have you given money to an environmental group? (GRNMONEY)

	Freq.	%
■ 1) YES	269	22.8
░ 2) NO	910	77.2
TOTAL (N)	1179	100.0
Missing	1638	

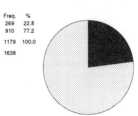

More than one-fifth of the GSS sample, and presumably of Americans, have donated funds to environmental groups.

What affects the likelihood of someone deciding to join an environmental group? While we can't look here at every possible factor for each area, we can examine a few. As noted earlier, education is often an important factor in decisions to join voluntary organizations. Let's see whether more educated people are more likely than those with less education to join an environmental group.

Data File: **GSS**

➤ Task: **Cross-tabulation**

➤ Row Variable: **166) GREEN GRP?**

➤ Column Variable: **15) EDUCATION**

➤ View: **Tables**

➤ Display: **Column %**

GREEN GRP? by EDUCATION
Cramer's V: 0.089 *

		EDUCATION					
		NOT H.S.	H.S. GRAD	SOME COLL.	COLL. GRAD	Missing	TOTAL
GREEN GRP?	YES	14	24	28	39	0	105
		6.2%	6.8%	8.6%	12.7%		8.7%
	NO	211	330	296	269	2	1106
		93.8%	93.2%	91.4%	87.3%		91.3%
	Missing	262	469	467	396	10	1604
	TOTAL	225	354	324	308	12	1211
		100.0%	100.0%	100.0%	100.0%		

Although the difference is small, people with a college degree are slightly more likely (12.7%) than those lacking a high school degree (6.2%) to join an environmental group (V = .09*).

Does education also predict the likelihood of donating money to an environmental organization?

Data File: **GSS**

Task: **Cross-tabulation**

➤ Row Variable: **167) GREEN MONY**

➤ Column Variable: **15) EDUCATION**

➤ View: **Tables**

➤ Display: **Column %**

GREEN MONY by EDUCATION
Cramer's V: 0.214 **

		EDUCATION					
		NOT H.S.	H.S. GRAD	SOME COLL.	COLL. GRAD	Missing	TOTAL
GREEN MONY	YES	20	61	80	107	1	268
		9.6%	17.5%	25.4%	35.1%		22.8%
	NO	189	288	235	198	0	910
		90.4%	82.5%	74.6%	64.9%		77.2%
	Missing	278	474	476	399	11	1638
	TOTAL	209	349	315	305	12	1178
		100.0%	100.0%	100.0%	100.0%		

This relationship is much stronger: people with a college degree are more likely (35.1%) than those without a high school degree (9.6%) to donate to an environmental group (V = .21**).

EXERCISE

4

REVIEW QUESTIONS

Based on the first part of this exercise, answer True or False to the following items:

Membership in sports and recreation groups is highest in European nations.	T F
Eastern European nations rank relatively low in the percent who say the family is very important in their lives.	T F
In the United States, Boy Scout membership seems higher in the South than in the North.	T F
Education makes no difference in decisions by Americans to donate money to environmental organizations.	T F
Confidence in organized labor is higher now than it was in the early 1970s.	T F

EXPLORIT QUESTIONS

1. In the preliminary section of this exercise, we saw that people with more education were more likely than those with less education to donate money to environmental organizations. In these next few questions you will explore whether certain other factors also predict the likelihood of donating money to these groups.

> ➤ *Data File:* **GSS**
> ➤ *Task:* **Cross-tabulation**
> ➤ *Row Variable:* **167) GREEN MONY**
> ➤ *Column Variable:* **26) FAM INCOME**
> ➤ *View:* **Tables**
> ➤ *Display:* **Column %**

a. What percent of people with high incomes ($50,000 or more) gave money to an environmental group? _____%

b. What percent of people with low incomes (under $25,000) gave money? _____%

c. Is V statistically significant?   Yes   No

d. Circle one of the following:

1. High-income people are more likely than low-income people to give money to an environmental group.

2. High-income people are less likely to give money.

3. Income is not related to the likelihood of donating money to these groups.

e. Do the results of this cross-tabulation illustrate the sociological perspective?

    1. Yes, because they suggest that income as one aspect of our social backgrounds influences one of our behaviors.

    2. No, because they show no relationship between family income and giving money to environmental organizations.

    3. No, because individuals are totally free to decide whether or not they want to give money to environmental organizations.

2. Should the region of the country where a person lives make a difference? What would you predict here?

> Data File: **GSS**
> Task: **Cross-tabulation**
> Row Variable: **167) GREEN MONY**
> ➤ Column Variable: **18) REGION**
> ➤ View: **Tables**
> ➤ Display: **Column %**

a. What percent of people in the Midwest have donated money to environmental groups?

    _____%

b. What percent of people from the West have donated money?

    _____%

c. Is V statistically significant?

    Yes   No

d. Circle one of the following:

    1. Taking into account statistical significance, Midwesterners are more likely than Westerners to give money to environmental groups.

    2. Taking into account statistical significance, Westerners are more likely than Midwesterners to give money.

    3. Taking into account statistical significance, region is not related to giving money to environmental groups.

e. Taking into account statistical significance, which of the following is true?

    1. The results of this cross-tabulation demonstrate that region influences whether people give money to environmental organizations.

    2. The results of this cross-tabulation demonstrate that region is not related to whether people give money to environmental organizations.

3. Next, consider the effects of race, looking at African Americans and whites.

> Data File: **GSS**
> Task: **Cross-tabulation**
> Row Variable: **167) GREEN MONY**
> ➤ Column Variable: **20) RACE**
> ➤ View: **Tables**
> ➤ Display: **Column %**

a. What percent of whites have given money to environmental groups?   _____%

b. What percent of African Americans have given money to environmental groups?   _____%

c. Is V statistically significant?   Yes   No

d. Circle one of the following:
    1. Whites are more likely than African Americans to give money to environmental groups.
    2. African Americans are more likely than whites to give money.
    3. Race is not related to giving money to environmental groups.

e. Do the results of this cross-tabulation illustrate the sociological perspective?
    1. Yes, because they suggest that race as one aspect of our social backgrounds influences one of our behaviors.
    2. No, because they show no relationship between race and giving money to environmental organizations.
    3. No, because individuals are totally free to decide whether or not they want to give money to environmental organizations.

4. Now select another GSS variable that might be related to whether respondents give money to environmental groups and obtain a cross-tabulation where 167) GREEN MONY is the row variable and the variable you select is the column variable.

   a. What variable did you select?   _____

   b. What was the value of V?   V = _____

   c. Was V statistically significant?   Yes   No

   d. Summarize the relationship that you found in your cross-tabulation.

5. Drawing on the information in Questions 1–3, indicate which one of the following descriptions best summarizes the characteristics of people who are most likely to give money to environmental organizations.
    1. Wealthy, white, South
    2. Poor, white, West
    3. Wealthy, black, Midwest
    4. Wealthy, white, Midwest

6. Does this description overall illustrate the sociological perspective?

   1. Yes, because it suggests that our social backgrounds influence one of our behaviors.

   2. No, because it shows that whether people give money to an environmental organization is purely a matter of individual choice.

7. One type of formal organization is the coercive organization, which, as the name implies, includes people who are forced to "belong to" the organization. One familiar type of coercive organization is prison. Your STATES data set lists the number of prisoners per 100,000 population in each state.

   > *Data File:* **STATES**
   > *Task:* **Mapping**
   > *Variable 1:* **63) PRISON**
   > *View:* **Map**

   a. Which two regions generally have the highest rates of incarceration (being behind bars)? (Circle one.)

      Northeast and Midwest

      South and West

      South and Midwest

   b. Which one of the following seems to be the **least** possible explanation for the regional pattern of imprisonment rates that you just found?

      1. Some regions have higher crime rates than other regions.

      2. The Midwest has an unusually high number of prisons.

      3. The chances of going to prison after arrest may be higher in some regions than in other regions.

   c. Suppose you have an adequate measure of regional differences in crime and want to determine whether such differences account for the regional differences in imprisonment. You obtain a scatterplot demonstrating the relationship between these two variables. If regional variation in crime does account for regional variation in imprisonment, what should the scatterplot look like?

      1. The higher the crime rate, the lower the imprisonment rate.

      2. The higher the crime rate, the higher the imprisonment rate.

      3. The lower the crime rate, the higher the imprisonment rate.

8. Earlier we looked at membership in organized labor. Historically, organized labor has found it difficult to gain a foothold in the South. Our hypothesis is that Southerners are less likely than people in other regions to belong to unions.

   > *Data File:* **GSS**
   > *Task:* **Cross-tabulation**
   > *Row Variable:* **86) UNIONIZED?**
   > *Column Variable:* **18) REGION**
   > *View:* **Tables**
   > *Display:* **Column %**

Discovering Sociology

a. Which region in the table has the lowest membership in labor
   unions? (Circle one.)

Northeast

South

Midwest

West

b. Taking V and statistical significance into account, are the people in the region you
   indicated above less likely than people in other regions to belong to labor unions?   Yes   No

9. Based on the results of the preceding example, do you think Southerners have less confidence in
   organized labor than people in other regions? Let's see.

>   Data File: **GSS**
>   Task: **Cross-tabulation**
>   ➤ Row Variable: **72) LABOR?**
>   ➤ Column Variable: **18) REGION**
>   ➤ View: **Tables**
>   ➤ Display: **Column %**

a. In which region are residents **most** likely to say they have "hardly any" confidence
   in organized labor? (Circle one.)

East

Midwest

South

West

b. Is V statistically significant?   Yes   No

10. Now select another GSS variable that might be related to the confidence people have in organized
    labor. Obtain a cross-tabulation where 72) LABOR? is the row variable and the variable you select is
    the column variable.

    a. What variable did you select?   _____

    b. Briefly state your hypothesis and explain why you think your hypothesis should be true.

    c. What was the value of V?   V = _____

    d. Was V statistically significant?   Yes   No

e.   Summarize the relationship that you found in your cross-tabulation.

DEVIANCE, CRIME, AND SOCIAL CONTROL

Tasks: Mapping, Scatterplot, Univariate, Historical Trends, Cross-tabulation
Data Files: NATIONS, STATES, GSS, HISTORY

If all societies have norms, or standards guiding behavior, it is also true that all societies have deviance, or violations of these norms. Emile Durkheim wrote long ago that a society without deviance is impossible, because there will always be some people who sometimes violate norms. The collective conscience, said Durkheim, is never strong enough to prevent all norm violation. If it were that strong, he added, the society would be very stagnant, because this would also be a society where people were unable to think creatively. Creative thinking presupposes a society where the collective conscience is weak enough that people are able to be independent thinkers. Yet such a society will also have people who choose to violate norms. You cannot have the first type of society without also having the second.

Deviance takes many forms. Some involves violations of norms that, in the grand scheme of things, are not that serious, for example dyeing your hair green or jaywalking, whereas other kinds of deviance involve much more important norms, such as murder, rape, and robbery. The latter types typically involve violations of formal, written norms, or criminal laws, and thus are more commonly called crimes.

This brief description might imply that something automatically is or is not deviant, but sociologists emphasize that what is considered deviant often depends on the reactions of others and the circumstances in which any particular act occurs. In this context, consider killing. If you kill someone because you're angry at him or her, you've committed a homicide, perhaps our most serious "street crime." But if you kill someone on the battlefield, you're instead doing your duty for your country, and if you kill many people on the battlefield, you may be considered a hero. In each case, killing has occurred; in the first case you go to prison or may even be executed, whereas in the second case you may get a medal.

This chapter examines some of the correlates of crime and deviance in the United States and around the world. It also explores the correlates of some attitudes toward these behaviors and the punishment of offenders who commit them. The chapter will help us once again to see what influence, if any, social backgrounds have on behaviors and attitudes.

A CROSS-CULTURAL LOOK AT DEVIANCE

As we look around the globe, we see much variation in approval and disapproval of various kinds of behaviors often considered deviant. Our NATIONS data set contains measures of views on many of these behaviors. Let's look at international opinion on suicide by using a measure of the percentage who say suicide is never acceptable.

SUICIDE NO -- PERCENT WHO THINK SUICIDE IS NEVER OK (WVS, 1997)

> ➤ *Data File:* **NATIONS**
> ➤ *Task:* **Mapping**
> ➤ *Variable 1:* **99) SUICIDE NO**
> ➤ *View:* **Map**

The darker the color, the greater the disapproval of suicide. Notice that disapproval appears highest in the Southern Hemisphere and some parts of Europe and lowest in other parts of Europe.

> *Data File:* **NATIONS**
> *Task:* **Mapping**
> *Variable 1:* **99) SUICIDE NO**
> ➤ *View:* **List: Rank**

RANK	CASE NAME	VALUE
1	Bangladesh	97.9
2	Brazil	87.4
3	Macedonia	83.5
4	Venezuela	83.3
5	Colombia	82.1
6	Ghana	79.7
7	Moldova	78.9
8	Nigeria	77.8
9	Georgia	76.7
10	Dominican Republic	75.8

Disapproval is highest in Bangladesh, where almost 98 percent think suicide is never acceptable. In contrast, only 26 percent of people in Sweden feel this way. In the United States, 60 percent think suicide is never acceptable.

What accounts for these international differences? One possible answer might lie in education: perhaps nations with lower levels of education are more likely to hold traditional attitudes toward suicide, and perhaps nations with higher levels of education are less likely. If so, we would expect a negative correlation between education and disapproval of suicide.

> *Data File:* **NATIONS**
> ➤ *Task:* **Scatterplot**
> ➤ *Dependent Variable:* **99) SUICIDE NO**
> ➤ *Independent Variable:* **116) EDUCATION**
> ➤ *Display:* **Reg. Line**

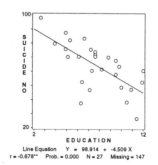

Line Equation   Y = 98.914 + -4.509 X
r = -0.678**   Prob. = 0.000   N = 27   Missing = 147

Discovering Sociology

As predicted, the higher the education level of a nation, the lower its disapproval of suicide (r = −.68**).

Let's turn now from international opinions on suicide to international differences in actual suicide rates.

Data File: **NATIONS**
➤ Task: **Mapping**
➤ Variable 1: **98) SUICIDE**
➤ View: **Map**

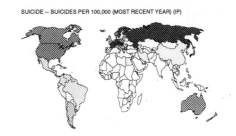

SUICIDE -- SUICIDES PER 100,000 (MOST RECENT YEAR) (IP)

Suicide rates tend to be highest in Europe.

Data File: **NATIONS**
Task: **Mapping**
Variable 1: **98) SUICIDE**
➤ View: **List: Rank**

RANK	CASE NAME	VALUE
1	Hungary	42.6
2	Sri Lanka	30.0
3	Finland	29.9
4	Denmark	26.6
5	Austria	23.1
6	Belgium	22.3
7	Switzerland	21.0
8	Russia	20.8
9	France	19.8
10	Luxembourg	18.6

Hungary leads the world with a suicide rate of 42.6 per 100,000 population. At 0.2 per 100,000, Iran and Syria have the lowest rate. The U.S. rate is 11.7.

Should more religious nations have lower suicide rates than less religious nations?

Data File: **NATIONS**
➤ Task: **Scatterplot**
➤ Dependent Variable: **98) SUICIDE**
➤ Independent Variable: **73) GOD IMPORT**
➤ Display: **Reg. Line**

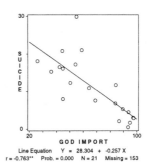

GOD IMPORT
Line Equation   Y = 28.304 + -0.257 X
r = -0.763**   Prob. = 0.000   N = 21   Missing = 153

Religious nations definitely have lower suicide rates than less religious nations (r = –0.76**). Religiosity is strongly related to suicide rates around the world.

CRIME AND DEVIANCE IN THE UNITED STATES

One of the most important questions in the United States is why people commit crime and deviance. Every year the FBI reports the amount and rate of various types of crime, most notably homicide, rape, aggravated assault, robbery, burglary, larceny, and motor vehicle theft. This report is called the *Uniform Crime Reports* (UCR). By convention, the number of crimes in a state is divided by the state's population and then multiplied by 100,000 to yield a crime rate per 100,000 population. This allows us to compare crime rates among states with very different populations. Let's examine some state-level correlates of violent crime (homicide, rape, aggravated assault, and robbery) in the United States to get a sociological sense of why people commit such crimes. We'll first map violent crime.

> ➤ Data File: **STATES**
> ➤ Task: **Mapping**
> ➤ Variable 1: **54) V.CRIME**
> ➤ View: **Map**

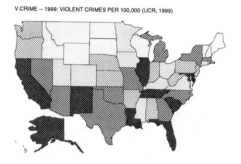

V.CRIME -- 1999: VIOLENT CRIMES PER 100,000 (UCR, 1999)

Notice that violent-crime rates are not randomly distributed throughout the United States. Instead, they are patterned geographically: the rates are generally highest in the South, the Far West, and a few northeastern states, and lowest in the upper Midwest and northern New England. Why does this pattern exist?

One possible explanation lies in whether a state is urban or rural. Urban areas obviously have a lot of people living very close together. These crowded conditions can prompt violent crime because they offer a ready supply of potential victims for would-be criminals and may also cause tempers to flare. Let's see whether the more urban states have higher violent-crime rates.

> Data File: **STATES**
> ➤ Task: **Scatterplot**
> ➤ Dependent Variable: **54) V.CRIME**
> ➤ Independent Variable: **19) %URBAN**
> ➤ Display: **Reg. Line**

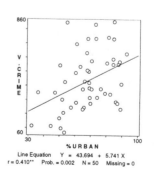

Line Equation   Y = 43.694 + 5.741 X
r = 0.410**   Prob. = 0.002   N = 50   Missing = 0

Our hypothesis is supported: the more urban a state, the higher its crime rate (r = .41**).

Another possible explanation comes from the *social disorganization* theory, which attributes high community crime rates to the weakening of social institutions such as religion and family by severe poverty, overcrowding, and other problems. One factor often cited here is the single-parent household. Areas with high levels of such households are apt to have higher crime rates for several reasons. First, they have fewer networks of "informal" social control and thus are less able than other areas to supervise adolescents and to be on the lookout for strangers. Second, areas with many single-parent households are more vulnerable to criminal victimization because single adults are more likely to be alone when outside, whether going to work or engaging in leisure-time activity, and are thus more vulnerable to personal crimes such as robbery, rape, and murder. Third, and this is the subject of much debate, single-parent households may do a poorer job of raising their children, who are thus more likely to become criminals when they grow up.

With these reasons in mind, let's see whether the states with the highest violent-crime rates also tend to be the states with the highest levels of single-parent households. Since most such households are headed by women, we'll use a measure of the percent of households that are female-headed with children present.

Data File: **STATES**
Task: **Scatterplot**
Dependent Variable: **54) V.CRIME**
➤ Independent Variable: **26) F HEAD/C**
➤ Display: **Reg. Line**

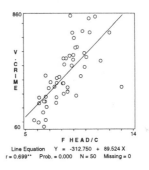

States with higher rates of single-parent households have higher violent-crime rates (r = .70**).

Now let's switch to the GSS data set to consider U.S. public opinion on crime and deviance issues. We'll start with a GSS question as to whether marijuana use should be made legal.

➤ Data File: **GSS**
➤ Task: **Univariate**
➤ Primary Variable: **50) GRASS?**
➤ View: **Pie**

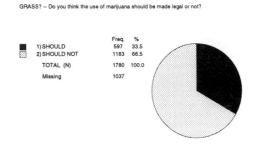

About one-third of the GSS respondents think marijuana use should be legalized.

> ➤ *Data File:* **HISTORY**
>   ➤ *Task:* **Historical Trends**
> ➤ *Variable:* **4) GRASS?**

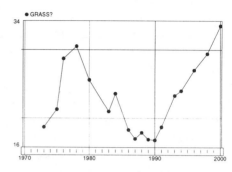

Percent saying marijuana should be made legal

This percent has certainly changed during the past quarter century. It rose through the 1970s, declined steadily until the late 1980s, then rose again during the 1990s.

Earlier we saw that the more-educated nations were more likely to accept suicide. Let's see whether the more-educated Americans are more likely to think marijuana use should be legalized. Do you think we'll find support for marijuana legalization rising with education?

> ➤ *Data File:* **GSS**
>   ➤ *Task:* **Cross-tabulation**
> ➤ *Row Variable:* **50) GRASS?**
> ➤ *Column Variable:* **15) EDUCATION**
>   ➤ *View:* **Tables**
>   ➤ *Display:* **Column %**

GRASS?   by   EDUCATION
Cramer's V: 0.066

		EDUCATION					
		NOT H.S.	H.S. GRAD	SOME COLL	COLL. GRAD	Missing	TOTAL
GRASS?	SHOULD	84	171	164	174	4	593
		29.1%	33.1%	32.2%	38.3%		33.5%
	SHOULD NOT	205	346	346	280	6	1177
		70.9%	66.9%	67.8%	61.7%		66.5%
	Missing	198	306	281	250	2	1037
	TOTAL	289	517	510	454	12	1770
		100.0%	100.0%	100.0%	100.0%		

Taking into account statistical significance, education is not related to support for marijuana legalization (V = .07). We cannot assume that support for marijuana legalization rises with education.

What about religiosity? Do you think we'll find that less religious people are more likely than more religious people to favor the legalization of marijuana?

Discovering Sociology

Data File: **GSS**
Task: **Cross-tabulation**
Row Variable: **50) GRASS?**
➤ Column Variable: **52) ATTEND**
➤ View: **Tables**
➤ Display: **Column %**

GRASS?   by   ATTEND
Cramer's V: 0.251 **

		ATTEND				
		NEVER	MONTH/YRLY	WEEKLY	Missing	TOTAL
GRASS?	SHOULD	169	324	84	20	577
		48.0%	38.0%	16.4%		33.6%
	SHOULD NOT	183	529	428	43	1140
		52.0%	62.0%	83.6%		66.4%
	Missing	231	487	302	17	1037
	TOTAL	352	853	512	80	1717
		100.0%	100.0%	100.0%		

People who never attend religious services are more likely (48.0 percent) than those who attend weekly (16.4 percent) to think that marijuana use should be legalized (V = .25**).

One reason crime gets so much attention these days is that so many people are concerned about it. The GSS asks, "Is there any area right around here—that is, within a mile—where you would be afraid to walk alone at night?" This is a common measure of fear of crime in the criminological literature. Let's see how many Americans are afraid of crime in their neighborhood.

Data File: **GSS**
➤ Task: **Univariate**
➤ Primary Variable: **168) FEAR WALK**
➤ View: **Pie**

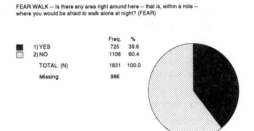

FEAR WALK -- Is there any area right around here -- that is, within a mile -- where you would be afraid to walk alone at night? (FEAR)

		Freq.	%
■	1) YES	725	39.6
▨	2) NO	1106	60.4
	TOTAL (N)	1831	100.0
	Missing	986	

About 40 percent say they'd be afraid to walk near their homes at night.

Has this percentage changed during the past 30 years?

➤ Data File: **HISTORY**
➤ Task: **Historical Trends**
➤ Variable: **5) FEAR WALK**

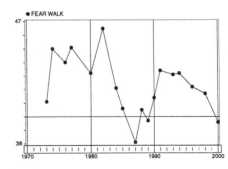

Percent saying they'd be afraid to walk near home at night

Fear of crime declined during the 1980s but is almost back to about its 1970s level.

Many aspects of our social backgrounds might affect the degree to which we're afraid of crime. Let's look first at gender. Do you think women will be more afraid than men or men more afraid than women, or do you think no gender difference exists?

> Data File: **GSS**
> Task: **Cross-tabulation**
> Row Variable: **168) FEAR WALK**
> Column Variable: **19) GENDER**
> View: **Tables**
> Display: **Column %**

FEAR WALK   by   GENDER
Cramer's V: 0.298 **

		GENDER		
		FEMALE	MALE	TOTAL
FEAR WALK	YES	537	188	725
		52.5%	23.2%	39.6%
	NO	485	621	1106
		47.5%	76.8%	60.4%
	Missing	566	420	986
	TOTAL	1022	809	1831
		100.0%	100.0%	

Women are more likely (52.5 percent) than men (23.2 percent) to say they'd be afraid to walk alone at night in their neighborhood. The relationship is very strong (V = .30**).

Now let's look at race. Most scholars agree that African Americans have higher rates than whites of crime and victimization. We can thus hypothesize that they should be more likely than whites to fear crime in their neighborhood.

Data File: **GSS**
Task: **Cross-tabulation**
Row Variable: **168) FEAR WALK**
> Column Variable: **20) RACE**
> View: **Tables**
> Display: **Column %**

FEAR WALK   by   RACE
Cramer's V: 0.058 *

		RACE			
		BLACK	WHITE	Missing	TOTAL
FEAR WALK	YES	125	556	44	681
		46.0%	38.1%		39.3%
	NO	147	903	56	1050
		54.0%	61.9%		60.7%
	Missing	160	779	47	986
	TOTAL	272	1459	147	1731
		100.0%	100.0%		

African Americans are more likely (46 percent) than whites (38.1 percent) to be afraid to walk alone at night (V = .06*).

Since women are so much more afraid of crime than men, let's hypothesize that they should be more likely than men to favor additional government spending on "halting the rising crime rate."

Data File: **GSS**

Task: **Cross-tabulation**

➤ Row Variable: **35) CRIME $**

➤ Column Variable: **19) GENDER**

➤ View: **Tables**

➤ Display: **Column %**

CRIME $ by   GENDER
Cramer's V: 0.097 **

		GENDER		
		FEMALE	MALE	TOTAL
CRIME $	TOO LITTLE	489	338	827
		64.9%	56.0%	60.9%
	RIGHT	231	223	454
		30.7%	36.9%	33.5%
	TOO MUCH	33	43	76
		4.4%	7.1%	5.6%
	Missing	835	625	1460
	TOTAL	753	604	1357
		100.0%	100.0%	

Although the difference is small, the hypothesis is supported. Women are slightly more likely (64.9 percent) than men (56 percent) to think the government is spending "too little" on crime control (V = .10**).

Since African Americans are more afraid of crime than whites, should we hypothesize that they should also be more likely than whites to favor additional spending on crime control?

Data File: **GSS**

Task: **Cross-tabulation**

Row Variable: **35) CRIME $**

➤ Column Variable: **20) RACE**

➤ View: **Tables**

➤ Display: **Column %**

CRIME $ by   RACE
Cramer's V: 0.115 **

		RACE			
		BLACK	WHITE	Missing	TOTAL
CRIME $	TOO LITTLE	150	639	38	789
		74.3%	58.8%		61.2%
	RIGHT	44	383	27	427
		21.8%	35.2%		33.1%
	TOO MUCH	8	65	3	73
		4.0%	6.0%		5.7%
	Missing	230	1151	79	1460
	TOTAL	202	1087	147	1289
		100.0%	100.0%		

African Americans are indeed more likely (74.3 percent) than whites (58.8 percent) to think the government is spending "too little" on crime (V = .12**).

WORKSHEET

NAME:

COURSE:

DATE:

EXERCISE

5

REVIEW QUESTIONS

Based on the first part of this exercise, answer True or False to the following items:

At the international level, the higher the education, the lower the disapproval
of suicide.   T   F

At the international level, the higher the religiosity, the lower the suicide rate.   T   F

In the United States, violent crime is higher in the Northeast than in any other region
of the country.   T   F

In the United States, more urban states have higher crime rates than rural states.   T   F

In the GSS, support for the legalization of marijuana has risen steadily since the
early 1970s.   T   F

EXPLORIT QUESTIONS

1. In the area of deviance and crime, gun control and the death penalty are two of the most controversial issues in the United States today. Let's start with gun control.

> ➤ Data File: **GSS**
> ➤ Task: **Cross-tabulation**
> ➤ Row Variable: **48) GUN LAW?**
> ➤ Column Variable: **19) GENDER**
> ➤ View: **Tables**
> ➤ Display: **Column %**

a. Who is more likely to favor gun control?   Women   Men

b. What would be a sociological explanation for the gender difference you found?

2. Does race affect support for gun control legislation?

> Data File: **GSS**
> Task: **Cross-tabulation**
> Row Variable: **48) GUN LAW?**
> ➤ Column Variable: **20) RACE**
> ➤ View: **Tables**
> ➤ Display: **Column %**

a. Taking into account statistical significance, are there racial differences in the United States in support for gun control?
        Yes   No

b. Which of the following statements best describes the results of this cross-tabulation?
   1. Blacks are less likely than whites to favor gun control as measured by the dependent variable.
   2. Whites are more likely than blacks to favor gun control.
   3. Blacks are more likely than whites to favor gun control.

3. Recall that women and African Americans are more afraid than men and whites respectively to walk around their neighborhoods at night. Your results for Questions 1 and 2 might thus suggest that fear of crime might be linked more generally to support for gun control.

> Data File: **GSS**
> Task: **Cross-tabulation**
> Row Variable: **48) GUN LAW?**
> ➤ Column Variable: **168) FEAR WALK**
> ➤ View: **Tables**
> ➤ Display: **Column %**

a. Is V statistically significant?
        Yes   No

b. What percent of people who are afraid to walk around their neighborhoods at night say they favor the type of gun law indicated in the variable?
    _____%

c. What percent of people who are not afraid to walk around their neighborhoods say they favor this gun law?
    _____%

d. Subtract the percentages from parts b and c and fill in the blank below.

People who are afraid to walk around their neighborhoods at night are _____ percent more likely than those who are not afraid to favor requiring a person to obtain a police permit before buying a gun.

4.   We turn now to the death penalty.

> Data File: **GSS**
>
> Task: **Cross-tabulation**
>
> ➤ Row Variable: **47) EXECUTE?**
>
> ➤ Column Variable: **19) GENDER**
>
> ➤ View: **Tables**
>
> ➤ Display: **Column %**

   a.   Is there a gender difference in support for the death penalty?   Yes   No

   b.   If so (and it might not be so), who favors the death penalty more?   Women

                                                              Men

                                                           Neither

   c.   What would be a sociological explanation for the result you find in the table?

5.   Should race affect support for the death penalty?

> Data File: **GSS**
>
> Task: **Cross-tabulation**
>
> Row Variable: **47) EXECUTE?**
>
> ➤ Column Variable: **20) RACE**
>
> ➤ View: **Tables**
>
> ➤ Display: **Column %**

   a.   African Americans are more likely than whites to support the death penalty.   T   F

   b.   More than half of all African Americans support the death penalty.   T   F

   c.   About 74 percent of whites support the death penalty.   T   F

   d.   What would be a sociological explanation for the result you find in the table?

6. In explaining the strong support by whites for the death penalty, some scholars say that racial prejudice plays a role. In this way of thinking, whites perceive that African Americans commit a disproportionately high number of homicides and other violent crimes. Whites who are racially prejudiced may thus favor the death penalty in part because they are prejudiced against blacks. We cannot test this perspective directly with our data, but we can at least determine whether, among whites, racial prejudice is indeed linked to support for the death penalty. The hypothesis is that whites who are more racially prejudiced should be more likely than whites who are less prejudiced to favor the death penalty. For our measure of racial prejudice, we'll use a variable that asks whites how they would feel if a close relative married an African American.

> Data File: **GSS**
> Task: **Cross-tabulation**
> Row Variable: **47) EXECUTE?**
> ➤ Column Variable: **121) MARRY BLK**
> ➤ Subset Variable: **20) RACE**
> ➤ Subset Category: **Include: 2) WHITE**
> ➤ View: **Tables**
> ➤ Display: **Column %**

   a. What percent of those who oppose a close relative marrying a black person favor the death penalty? _____%

   b. What percent of those who neither favor nor oppose this type of marriage favor the death penalty? _____%

   c. What percent of those who favor this type of marriage favor the death penalty? _____%

   d. Judging from V, can we conclude that racial prejudice among whites is related to support for the death penalty?   Yes   No

7. Now let's go back to the NATIONS data set.

> ➤ Data File: **NATIONS**
> ➤ Task: **Mapping**
> ➤ Variable 1: **96) PROSTITUTE**
> ➤ View: **List: Rank**

   a. Which country has the highest percent who think prostitution is never OK? _____

   b. Which country has the lowest percent? _____

   c. What is the percent for the United States? _____%

d. Fill in the percentages for the following nations:

Chile	_____	%
Sweden	_____	%
United Kingdom	_____	%
Mexico	_____	%

8. Should disapproval of prostitution be higher in the more religious nations?

> *Data File:* **NATIONS**
> ➤ *Task:* **Scatterplot**
> ➤ *Dependent Variable:* **96) PROSTITUTE**
> ➤ *Independent Variable:* **74) PRAY?**
> ➤ *Display:* **Reg. Line**

a. What is the value of r?   r = _____

b. Is r statistically significant?   Yes   No

c. If we hypothesize that the more religious nations should be more likely to think
    that prostitution is never OK, do the data support the hypothesis?   Yes   No

d. What other variable about a nation do you think should affect the degree to which its citizens oppose
    prostitution? Why?

9. In the preliminary section of this exercise, we saw a religiosity difference in the GSS in support for
    the legalization of marijuana, with less-religious people more in favor of legalization than more-reli-
    gious people. Will gender make a difference? Since males are socialized to be more daring and
    assertive, let's hypothesize that they'll be more likely than women to favor marijuana legalization.

> ➤ *Data File:* **GSS**
> ➤ *Task:* **Cross-tabulation**
> ➤ *Row Variable:* **50) GRASS?**
> ➤ *Column Variable:* **19) GENDER**
> ➤ *View:* **Tables**
> ➤ *Display:* **Column %**

a. What percent of men favor legalization?   _____

SOCIAL STRATIFICATION

Tasks: Mapping, Scatterplot, Cross-tabulation, Univariate
Data Files: NATIONS, STATES, GSS

In today's world, most societies are ***stratified***, meaning that they're characterized by an unequal distribution of resources that the society values. This unequal distribution yields a social hierarchy, with some people ranked at the top of the hierarchy, others ranked at the bottom, and the rest somewhere in between. As Max Weber pointed out, the resources in the modern world that most determine social ranking are wealth, power, and prestige. At the top of stratification systems today, then, are the people who have the most wealth, power, and/or prestige, while at the bottom are the people who have the least. Although the people with the most wealth usually also have much power and prestige and vice versa, this is not always true, as Weber also pointed out. For example, a big-time drug dealer might be very wealthy, but obviously has little prestige outside of the criminal world. Professional athletes are also quite wealthy, but again lack power in the political sense that Weber meant it.

Sociologists offer two basic explanations for why stratification exists. Briefly, *functional theory* assumes that society must promise higher incomes so that talented people will decide to get the education and training they need to enter the most important and highly skilled occupations. This automatically means that they will end up with higher incomes than other people, resulting in a stratified society. In this view, stratification is both necessary and inevitable. In contrast, *conflict theory* assumes that stratification results from the lack of equal opportunity for people born into poverty and for those facing discrimination because of their race and ethnicity. In this view, stratification is neither necessary nor inevitable.

Whatever its sources, social stratification has important consequences for ***life chances***, or the outcomes we can expect in life's endeavors. Some common negative outcomes include poor physical and mental health, low educational achievement, and divorce and other family problems. In all of these areas, according to much research, the poor fare far worse than the rich.

If some individuals are much wealthier than other individuals, it is also true that some nations are much wealthier than other nations. We call this difference in wealth ***global stratification***. The nations at the bottom of this social hierarchy are desperately poor, with hunger and disease rampant. Many of the differences in life chances between the poor and the nonpoor in individual societies also exist between poor and rich societies at the global level.

This chapter examines some key features of stratification throughout the world and within the United States. Although our primary focus will be the correlates, including several life chances, of both types of stratification, we will also look at views on such issues as government spending and attitudes toward the poor.

GLOBAL STRATIFICATION

Let's begin our look at global stratification by first getting a picture of where the richest and poorest nations lie, using a common measure of the annual gross domestic product per capita.

➤ *Data File:* **NATIONS**
➤ *Task:* **Mapping**
➤ *Variable 1:* **30) GDP/CAP**
➤ *View:* **Map**

GDP/CAP -- GROSS DOMESTIC PRODUCT PER CAPITA IN U.S. DOLLARS (HDR, 2001)

The darker the color, the higher the nation's annual national product per capita. The nations in North America and Western Europe tend to be the wealthiest on this measure, and nations in Africa and Asia the poorest. The nations of the world are, indeed, stratified.

Data File: **NATIONS**
Task: **Mapping**
➤ *Variable 1:* **10) INF. MORTL**
➤ *View:* **Map**

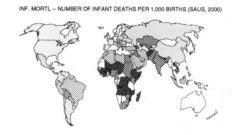

INF. MORTL -- NUMBER OF INFANT DEATHS PER 1,000 BIRTHS (SAUS, 2000)

Infant mortality is highest in Africa and parts of Asia.

Data File: **NATIONS**
Task: **Mapping**
Variable 1: **10) INF. MORTL**
➤ *View:* **List: Rank**

RANK	CASE NAME	VALUE
1	Angola	195.8
2	Afghanistan	149.3
3	Sierra Leone	148.7
4	Mozambique	139.9
5	Guinea	131.0
6	Somalia	125.8
7	Niger	124.9
8	Mali	123.3
9	Malawi	122.3
10	Rwanda	120.1

In the five nations with the highest infant mortality rates, more than 130, or 13 percent, of every 1,000 infants die before age 1. In the nations with the lowest rates, fewer than 5, or under 1 percent, of every 1,000 infants die before age 1.

Data File: **NATIONS**
➤ Task: **Scatterplot**
➤ Dependent Variable: **10) INF. MORTL**
➤ Independent Variable: **30) GDP/CAP**
➤ Display: **Reg. Line**

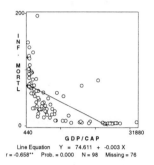

The lower the national wealth, the higher the infant mortality. A baby born in a very poor nation has a much greater chance of dying young (r = −.66**).

It's not a matter of life and death, but people in the United States take telephones, televisions, and cars for granted. These American "necessities of life" are, however, virtually unknown in many parts of the world. Let's use a scatterplot to show the strong association between national wealth and the presence of one of these items.

Data File: **NATIONS**
Task: **Scatterplot**
➤ Dependent Variable: **45) CELL PHONE**
➤ Independent Variable: **30) GDP/CAP**
➤ Display: **Reg. Line**

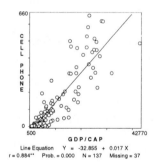

Wealthier nations have many more cell phones per 100,000 people than poorer nations. The correlation, r, is a very high .88**.

As this scatterplot indicates, what is considered a "necessity of life" in the United States and other wealthy nations is an unaffordable luxury in much of the rest of the world.

STRATIFICATION IN THE UNITED STATES

To begin our look at U.S. stratification, let's first see which states are the poorest. We'll use the percent of the state's population below the poverty level.

➤ *Data File:* **STATES**
➤ *Task:* **Mapping**
➤ *Variable 1:* **45) %POOR**
➤ *View:* **Map**

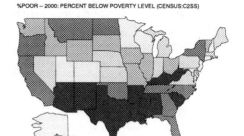

%POOR -- 2000: PERCENT BELOW POVERTY LEVEL (CENSUS:C2SS)

Generally, the South is the poorest region of the country.

Data File: **STATES**
Task: **Mapping**
Variable 1: **45) %POOR**
➤ *View:* **List: Rank**

RANK	CASE NAME	VALUE
1	Louisiana	20.0
2	West Virginia	19.3
3	Mississippi	18.2
4	New Mexico	18.0
5	Arkansas	17.4
6	Kentucky	16.5
7	Alabama	16.0
8	Arizona	15.6
9	Texas	15.3
10	South Carolina	14.8

The poverty rate ranges from a high of 20 percent in Louisiana to a low of 6 percent in New Hampshire. Louisiana's rate is three times times greater than New Hampshire's. Where does your home state rank?

What are the implications of the states' poverty levels for the life chances of their residents? Earlier we saw that global stratification is linked to international differences in infant mortality. Does a similar relationship hold true within the United States? Let's first see which states rank highest and lowest on this life chance.

Data File: **STATES**
Task: **Mapping**
➤ *Variable 1:* **34) INF.MORT.**
➤ *View:* **Map**

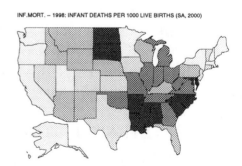

INF.MORT. – 1998: INFANT DEATHS PER 1000 LIVE BIRTHS (SA, 2000)

The Southeast and a few other states have the highest infant mortality rates.

Data File: **STATES**
➤ Task: **Scatterplot**
➤ Dependent Variable: **34) INF.MORT.**
➤ Independent Variable: **45) %POOR**
➤ Display: **Reg. Line**

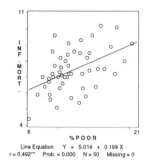

Line Equation   Y = 5.014 + 0.199 X
r = 0.492**   Prob. = 0.000   N = 50   Missing = 0

The poorer the state, the higher its infant mortality rate (r = .50**). Although the United States has a lower infant mortality rate than most other nations, poverty at the state level is still linked to the chances of an infant's dying before age 1.

Recall that the functional theory of stratification assumes we would not have enough talented people entering the most important and skilled occupations unless they expected higher incomes for doing so. The GSS asks respondents whether they agree that "no one would study for years to become a lawyer or doctor unless they expected to earn a lot more than ordinary workers." This question is a rough measure of this central belief of functional theory. How would you answer it?

➤ Data File: **GSS**
➤ Task: **Univariate**
➤ Primary Variable: **134) INEQUAL 4**
➤ View: **Pie**

INEQUAL 4 -- No one would study for years to become a lawyer or doctor unless they expected to earn a lot more than ordinary workers--do you agree or disagree?

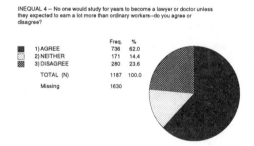

	Freq.	%
1) AGREE	736	62.0
2) NEITHER	171	14.4
3) DISAGREE	280	23.6
TOTAL (N)	1187	100.0
Missing	1630	

Almost two-thirds of Americans believe there would be too few lawyers and doctors if their incomes were not much higher than those of ordinary workers. By implication, a strong majority of the public accepts the central view of functional theory.

Does this acceptance differ by social class? Functional theory in the broader discipline of sociology assumes American society is characterized by shared values and opinions on important social issues. In contrast, conflict theory assumes that divisions of opinion instead exist on these issues, with groups of people holding beliefs according to their "vested interests" in maintaining their class position. In particular, wealthy people should be more likely than poorer people to hold views that support the existing system of social stratification.

This is an interesting and important theoretical debate. If functional theory is correct, acceptance of its central view on stratification should *not* differ by social class.

Data File:	**GSS**
➤ Task:	**Cross-tabulation**
➤ Row Variable:	**134) INEQUAL 4**
➤ Column Variable:	**26) FAM INCOME**
➤ View:	**Tables**
➤ Display:	**Column %**

INEQUAL 4   by   FAM INCOME
Cramer's V: 0.070 *

		FAM INCOME				
		$0K-24.9K	$25K-49.9K	$50K +	Missing	TOTAL
I N E Q U A L 4	AGREE	250	193	215	78	658
		65.6%	63.5%	56.7%		61.8%
	NEITHER	59	39	57	16	155
		15.5%	12.8%	15.0%		14.6%
	DISAGREE	72	72	107	29	251
		18.9%	23.7%	28.2%		23.6%
	Missing	511	419	462	238	1630
	TOTAL	381	304	379	361	1064
		100.0%	100.0%	100.0%		

The relationship is small, but poorer people are actually slightly more likely (65.6 percent) than wealthier people (56.7 percent) to agree that there would be too few doctors and lawyers if their incomes were not so high (V = .07*). Ironically, poorer people are more likely, then, to hold a belief that supports the existing system of social stratification.

We can also address this theoretical debate with a dependent variable that reflects a central belief of conflict theory: that inequality benefits the wealthy interests in society. The GSS asks respondents whether they agree that "inequality continues to exist because it benefits the rich and powerful." How would you answer this question?

Data File:	**GSS**
➤ Task:	**Univariate**
➤ Primary Variable:	**133) INEQUAL 3**
➤ View:	**Pie**

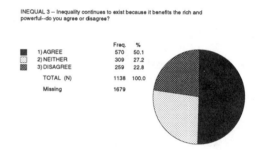

INEQUAL 3 -- Inequality continues to exist because it benefits the rich and powerful--do you agree or disagree?

		Freq.	%
■	1) AGREE	570	50.1
▨	2) NEITHER	309	27.2
▩	3) DISAGREE	259	22.8
	TOTAL (N)	1138	100.0
	Missing	1679	

Half the public agrees that inequality exists because it benefits the rich and powerful. Although this figure is not as large as the one we just found for acceptance of functional theory's central belief, it does indicate substantial support for an important view of conflict theory. Does this support differ by social class? Conflict theory would predict that poorer people should be more likely than wealthier people to blame inequality on the rich and powerful.

Discovering Sociology

Data File: **GSS**
➤ Task: **Cross-tabulation**
➤ Row Variable: **133) INEQUAL 3**
➤ Column Variable: **26) FAM INCOME**
➤ View: **Tables**
➤ Display: **Column %**

INEQUAL 3   by   FAM INCOME
Cramer's V: 0.078 *

		FAM INCOME				
		$0K-24.9K	$25K-49.9K	$50K +	Missing	TOTAL
INEQUAL 3	AGREE	204	140	171	55	515
		56.5%	47.3%	45.8%		50.0%
	NEITHER	92	88	101	28	281
		25.5%	29.7%	27.1%		27.3%
	DISAGREE	65	68	101	25	234
		18.0%	23.0%	27.1%		22.7%
	Missing	531	427	468	253	1679
	TOTAL	361	296	373	361	1030
		100.0%	100.0%	100.0%		

Again the relationship is small, but this time conflict theory is supported, as poorer people are more likely (56.5 percent) than wealthier people (45.8 percent) to believe that inequality exists because it benefits the rich and powerful (V = .08*). Though certainly not the last word on the subject, our results overall provide some support for both functional and conflict theories.

Now, let's examine the effects of social class on different life changes. The GSS includes several variables measuring different aspects of how well the respondents think their lives are going. Let's examine the relationship between family income and two of these variables.

One of the most important life chances is our health. The GSS asks respondents to rate their own health as excellent, good, fair, or poor.

Data File: **GSS**
Task: **Cross-tabulation**
➤ Row Variable: **65) HEALTH**
➤ Column Variable: **26) FAM INCOME**
➤ View: **Tables**
➤ Display: **Column %**

HEALTH   by   FAM INCOME
Cramer's V: 0.221 **

		FAM INCOME				
		$0K-24.9K	$25K-49.9K	$50K +	Missing	TOTAL
HEALTH	EXCELLENT	138	202	276	86	616
		18.4%	33.5%	40.5%		30.3%
	GOOD	333	309	335	126	977
		44.3%	51.2%	49.1%		48.0%
	FAIR/POOR	280	92	71	76	443
		37.3%	15.3%	10.4%		21.8%
	Missing	141	120	159	73	493
	TOTAL	751	603	682	361	2036
		100.0%	100.0%	100.0%		

Low-income people are less likely (18.4 percent) than high-income people (40.5 percent) to rate their health as excellent, and they are much more more likely to rate their health as only fair or poor (V = .22**). Social class certainly does seem to be related to our health.

Could the problems of low-income families lead to depression? The GSS asks how often during the past month respondents felt "downhearted and blue." Let's see whether low-income people are more likely to feel this way.

Data File: **GSS**
Task: **Cross-tabulation**
➤ Row Variable: **153) DOWN BLUE**
➤ Column Variable: **26) FAM INCOME**
➤ View: **Tables**
➤ Display: **Column %**

DOWN BLUE   by   FAM INCOME
Cramer's V: 0.137 **

		FAM INCOME				
		$0K-24.9K	$25K-49.9K	$50K +	Missing	TOTAL
DOWN BLUE	MOST	31	18	12	9	61
		6.8%	4.9%	3.0%		5.0%
	SOME	146	70	66	47	282
		32.2%	19.2%	16.3%		23.0%
	LTTLE/NONE	277	277	327	129	881
		61.0%	75.9%	80.7%		72.0%
	Missing	438	358	436	176	1408
	TOTAL	454	365	405	361	1224
		100.0%	100.0%	100.0%		

To make this table easier to interpret, add the percent in each column who said they felt downhearted and blue either most or some of the time. When you do so, you'll find that 39.0 percent of low-income people said most or some of the time, compared to only 19.3 percent of the high-income people. The low-income people are thus twice as likely as their high-income counterparts to have felt downhearted and blue during the past month (V = .14**). Judging from our results for health and feeling blue, low-income people in the United States certainly face more negative life chances.

Or do they? As Exercise 2 explained, the fact that two variables are statistically related does not mean one necessarily affects the other. It is possible that the statistical relationship exists only because we have not taken into account the effects of some other factor. To use a silly example to make the point, a survey would undoubtedly find that people who frequently listen to "punk rock" music are more likely to have bad acne than those who do not listen to this music. Does this relationship mean that listening to such music somehow causes people to break out in acne? Of course not! A much more plausible explanation is that young people are much more likely than older people to listen to punk rock and are also, for very different reasons, to have bad acne. If we "controlled for" age by looking at younger and older people separately, we would find no relationship between listening to punk rock and having acne within either age group. Because age affects the likelihood both of listening to punk rock and of having acne, we do need to control for it.

A statistical relationship that exists between an independent variable and a dependent variable only because we have not controlled for the effects of a third variable is said to be a *spurious relationship*. Any relationship between listening to punk rock and having acne is obviously spurious because we have not controlled for age.

Yet, even a more plausible relationship between an independent and dependent variable may turn out to be spurious once other variables have been controlled for. To return to our income and feeling-blue relationship, it is possible that even this very plausible relationship is spurious because we have not controlled for an important variable that may affect both income and the likelihood of feeling blue. One such variable is age. Elderly people are poorer on the average than younger people and may also, for different reasons such as health problems, be more likely to feel blue. Just as the relationship between punk rock and acne turns out to be spurious once age is controlled for, so might the relationship between income and feeling blue. To rule out this possibility of spuriousness, we need to re-examine the relationship between income and feeling blue while controlling for age.

To do this, we simply use ExplorIt's "Control Variables" option that is part of the CROSS-TABULATION procedure. As the following ExplorIt guide indicates, simply type in the number or name of 14) AGE 65+ as a control variable after you have typed in 153) DOWN BLUE and 26) FAM INCOME as the dependent (row) and independent (column) variables, respectively. Once you have done so, you will first see a screen that shows the relationship between income and feeling blue just for people ages 18–64. To show the relationship between income and feeling blue just for people ages 65+, simply click the right-facing button under the "control" option on the screen.

Data File: **GSS**
Task: **Cross-tabulation**
Row Variable: **153) DOWN BLUE**
Column Variable: **26) FAM INCOME**
➤ Control Variable: **14) AGE 65+**
➤ View: **Tables**
➤ Display: **Column %**

DOWN BLUE   by   FAM INCOME
Controls: AGE 65+:   18-64
Cramer's V: 0.132 **

		FAM INCOME				
		$0K-24.9K	$25K-49.9K	$50K +	Missing	TOTAL
DOWN BLUE	MOST	23	18	12	4	53
		6.9%	5.6%	3.2%		5.2%
	SOME	104	63	58	31	225
		31.4%	19.6%	15.5%		21.9%
	LTTLE/NONE	204	240	304	96	748
		61.6%	74.8%	81.3%		72.9%
	Missing	334	314	396	125	1169
	TOTAL	331	321	374	256	1026
		100.0%	100.0%	100.0%		

When you look first at the relationship for people ages 18–64, notice that it is still statistically significant: people in the lowest income bracket are more likely than those in the highest bracket to have felt blue most or some of the time during the past month (V = .13**). When you next look at the relationship for people ages 65 and older, a similar relationship again emerges (V = .17*). The relationship between income and feeling blue, then, is not spurious when we control for age. Such a failure to find a spurious relationship when we might have expected it gives us more confidence that there is, in fact, a real relationship between income and feeling blue, which is exactly what studies of poverty and psychological well-being have found.

REVIEW QUESTIONS

Based on the first part of this exercise, answer True or False to the following items:

The wealthiest nations in the world tend to be in North America and Western Europe.	T	F
Even in the poorest nations most people have cell phones.	T	F
The poorest region in the United States is the Midwest.	T	F
Despite what many people think, poverty and infant mortality in the United States are not related.	T	F
Less than half of the public believes that inequality benefits the rich and powerful.	T	F
Low-income people are more likely than high-income people to report bad health.	T	F

EXPLORIT QUESTIONS

1. As the preliminary section of this exercise noted, conflict theory assumes that people's beliefs to some extent reflect their *vested interests* stemming from their own class position. Thus, the rich should have a more negative view of the poor than the poor have of themselves and should be less likely than the poor to advocate policies designed to reduce poverty. Let's test these assumptions.

   The NATIONS data set includes a variable measuring the percent of a nation's population who said the reason that people in their country are poor is that "there is injustice in our society."

   > *Data File:* **NATIONS**
   > *Task:* **Mapping**
   > *Variable 1:* **89) INJUSTICE**
   > *View:* **List: Rank**

   a. What percent of the U.S. population subscribes to this belief? _____%

   b. Of the 43 nations where this question was asked, where does the United States rank? _____

   c. To explain this ranking, which of the following characteristics would conflict theory be most likely to emphasize?
      1. People in the United States watch a lot of TV.
      2. The United States is a very wealthy nation.
      3. There is not much injustice in the United States.

2. Let's see whether national differences in this belief about injustice and the poor are related to nation-
    al differences in wealth. If our discussion at the beginning of this exercise is correct, the wealthier the
    nation, the *less* likely its population should believe that poverty is due to injustice.

>   *Data File:* **NATIONS**
> ➤ *Task:* **Scatterplot**
> ➤ *Dependent Variable:* **89) INJUSTICE**
> ➤ *Independent Variable:* **30) GDP/CAP**
> ➤ *Display:* **Reg. Line**

a. Which of the following provides the best interpretation of this scatterplot?

    1. The wealthier the nation, the more likely its population believes that poverty is due to
        injustice.

    2. The poorer the nation, the less likely its population believes that poverty is due to injustice.

    3. The wealthier the nation, the less likely its population believes that poverty is due to
        injustice.

b. Do the results of the scatterplot support conflict theory's assumptions?

    1. Yes, because r is statistically significant, and the variables are related in the way that
        conflict theory would predict.

    2. No, because although r is statistically significant, the relationship between the variables is
        the opposite of what conflict theory would predict.

    3. No, because r is not statistically significant.

3. The GSS also asks respondents whether they agree that the government should reduce the differ-
    ences in income between people with high incomes and people with low incomes.

a. What would be conflict theory's prediction of how family income should be related to responses to
    this question?

    1. The higher the income, the greater the belief that the government should reduce
        income differences.

    2. The higher the income, the less the belief that the government should reduce income dif-
        ferences.

4. Use the results of this table to answer the following questions.

> ➤ *Data File:* **GSS**
> ➤ *Task:* **Cross-tabulation**
> ➤ *Row Variable:* **135) GOVT EQUAL**
> ➤ *Column Variable:* **26) FAM INCOME**
> ➤ *View:* **Tables**
> ➤ *Display:* **Column %**

a. Is this statement, "It is the responsibility of the government to reduce the differences in income between people with high incomes and those with low incomes" a belief of functional theory or of conflict theory?   functional theory   conflict theory

b. Which hypothesis below would conflict theory predict?

    1. People with higher incomes are more likely than those with lower incomes to believe that the government should reduce income differences.

    2. People with higher incomes are less likely than those with lower incomes to believe that the government should reduce income differences.

c. More than half of medium-income people think the government should try to reduce income differences.   T   F

d. This table provides support for the conflict theory.   T   F

5. Another GSS question asks whether large differences in income are necessary for American prosperity.

a. Why should this be considered a belief of functional theory?

b. What hypothesis would conflict theory predict between family-income and responses to this question?

6. Let's test this hypothesis.

>   *Data File:*   **GSS**
>   *Task:*   **Cross-tabulation**
> ➤   *Row Variable:*   **179) INEQUAL 5**
> ➤   *Column Variable:*   **26) FAM INCOME**
>   ➤   *View:*   **Tables**
>   ➤   *Display:*   **Column %**

a. Is V statistically significant?   Yes   No

b. Do these results of this hypothesis support conflict theory?

    1.   Yes, because V is statistically significant.

    2.   No, because although V is statistically significant, the relationship tends to be the opposite of what functional theory would predict.

    3.   No, because V is not statistically significant.

7.   In the preliminary section of this exercise, we saw that infant mortality is higher in the poorer nations. If infants sometimes die before age 1, mothers sometimes die during childbirth. A common indicator of this problem is maternal mortality, the number of maternal deaths per 100,000 live births. In the NATIONS data set, this rate ranges from 1,100 (Mozambique) to only 1 (Greece). Let's examine its relationship with national wealth.

> *Data File:* **NATIONS**
> *Task:* **Scatterplot**
> *Dependent Variable:* **11) MOM MORTAL**
> *Independent Variable:* **30) GDP/CAP**
> *Display:* **Reg. Line**

a.   What conclusion would you draw from this scatterplot?

    1.   The wealthier a nation, the higher its rate of maternal mortality.

    2.   The poorer a nation, the lower its rate of maternal mortality.

    3.   The poorer a nation, the higher its rate of maternal mortality.

b.   What is another conclusion that can be drawn from this scatterplot?

    1.   A nation's wealth or poverty has important implications for its people's life chances.

    2.   A nation's wealth or poverty has no implications for its people's life chances.

8.   Earlier we also saw that cell phones are much less common in the poorest nations. Do you think we'll find a similar trend in the United States for phones of all types when we compare the poorer and wealthier states? (Again, be sure to read the complete variable descriptions for the variables being used in your analysis.)

> *Data File:* **STATES**
> *Task:* **Scatterplot**
> *Dependent Variable:* **29) NO PHONE**
> *Independent Variable:* **45) %POOR**
> *Display:* **Reg. Line**

a.   This scatterplot clearly indicates that

    1.   poorer states are more likely to have housing units without telephones.

    2.   whether a state has a high or low rate of housing units without phones depends on which region of the country the state is in.

    3.   there is no relationship between the poverty level of a state and the percent of its housing units that lack telephones.

9. An important topic in the study of stratification is the issue of social mobility, or the degree to which people are able to move up the socioeconomic ladder. Since its inception, the United States has been thought to be a land of equal opportunity where people who work hard enough can pull themselves up by their bootstraps. Compared to most countries in the world, the United States does enjoy more social mobility. Still, many sociologists emphasize that the degree of achieving the American dream depends to a large degree on where one starts out in life. We can explore this point by comparing GSS respondents' financial circumstances when they were young to their own family's income as adults. In the table that follows, the independent variable indicates whether both of the respondent's parents were in low-prestige jobs (as determined by a national ranking of prestige scores assigned to a range of jobs) or high-prestige jobs. These categories roughly correspond to low-paying and high-paying jobs, respectively.

> ➤ *Data File:* **GSS**
>   ➤ *Task:* **Cross-tabulation**
> ➤ *Row Variable:* **26) FAM INCOME**
> ➤ *Column Variable:* **16) PARS.PRESG**
>   ➤ *View:* **Tables**
>   ➤ *Display:* **Column %**

  a. This table indicates that

      1. respondents whose parents were in high-prestige jobs are more likely than those whose parents were in low-prestige jobs to end up with higher family incomes themselves.

      2. respondents whose parents were in high-prestige jobs are more likely than those whose parents were in low-prestige jobs to end up with lower family incomes themselves.

      3. there is no relationship between the job prestige of respondents' parents and the current family income of the respondents.

  b. Do the results of this cross-tabulation indicate that where people start out in life
     is related to where they end up?   Yes   No

10. How much of an effect does your own education have on your annual income? Professors believe a student should be in college primarily to become a learned person, but that's a little idealistic. Although students often have this goal, they also hope that a college degree will improve their chances of a decent income. Let's see whether college pays off. The dependent variable will be the respondent's personal income, divided into three categories.

> *Data File:* **GSS**
>   *Task:* **Cross-tabulation**
> ➤ *Row Variable:* **27) OWN INCOME**
> ➤ *Column Variable:* **15) EDUCATION**
>   ➤ *View:* **Tables**
>   ➤ *Display:* **Column %**

a. Look at the row for the highest income category listed, $35,000 and up. Approximately how many times more likely is a college graduate to end up in the highest income bracket than is a person without a high school degree?   _____ times

b. Subtract the appropriate percents and complete this sentence: People without a high school degree are ____ percent more likely than those with a college degree to have annual incomes under $17,500.

c. Judging from the results of this cross-tabulation, does education seem important to the incomes we achieve?   Yes   No

11. Our social class is often related to our feelings or attitudes on many kinds of issues. Select a GSS variable dealing with an attitude or a feeling and use it as a dependent (row) variable in a cross-tabulation where 26) FAM INCOME is the independent (column) variable. Make sure you use the [Column %] option for the display. Then answer the following questions.

    a. What dependent variable did you select (number and name)?   _____

    b. Which hypothesis is this cross-tabulation testing?

    c. Was the V for your table statistically significant?   Yes   No

    d. Print out the percentaged results and turn them in with your assignment. On a separate sheet of paper, summarize the results of your analysis. Also indicate what conclusion relevant to your hypothesis you drew from your table. (Turn in this paper too.)

12. In the preliminary section of this exercise, we found that the relationship between family income and feeling blue was not spurious when we controlled for age. Another variable that we might want to control is race, the subject of the next exercise, as it is possible that the income–feeling blue relationship might turn out to be spurious once race is taken into account. In this way of thinking, African Americans are poorer than whites and might also, for different reasons such as racial slights, be more likely to feel blue. If we control for race and find out that the income–feeling blue relationship is not spurious, that again increases our confidence that this relationship is real.

> Data File: **GSS**
> Task: **Cross-tabulation**
> ➤ Row Variable: **153) DOWN BLUE**
> ➤ Column Variable: **26) FAM INCOME**
> ➤ Control Variable: **20) RACE**
> ➤ View: **Tables**
> ➤ Display: **Column %**

a. Is V statistically significant for the income–feeling blue relationship just among African Americans?   Yes   No

b. Is V statistically significant for the income–feeling blue relationship just among whites?   Yes   No

c. Do these results indicate that the original relationship between income and feeling blue is not spurious?   Yes   No

◆ EXERCISE 7 ◆
RACE AND ETHNICITY

Tasks: Mapping, Scatterplot, Univariate, Historical Trends, Cross-tabulation
Data Files: NATIONS, STATES, GSS, HISTORY

Perhaps the most important sociological aspect of race and ethnicity is that people of different races and ethnicities are treated differently. More precisely, some are treated better, while others are treated worse. The latter are said to belong to a racial or ethnic *minority group*, a category of people treated negatively because of their perceived physical or cultural characteristics. Negative outcomes for minority groups include housing and job discrimination, poverty and low education, and serious health problems. Minority groups in the United States and elsewhere have often been the victims of hate crimes, and, in other nations, have suffered virtual genocide.

This exercise explores several aspects of race and ethnicity in the United States and throughout the world. Our major focus will be on the extent and correlates of racial and ethnic prejudice and on the negative life chances of minority groups. We'll begin by exploring these topics on the international level and then switch our attention to the United States.

RACE AND ETHNICITY IN THE INTERNATIONAL ARENA

It's surprisingly difficult to measure how much racial and ethnic diversity exists within any single nation. Using a complex formula, sociologist Rodney Stark calculated the odds that any two persons in a given nation will differ in their race, religion, ethnicity or tribe, or language group. Let's use this measure of multiculturalism to see which regions of the world have the most and least diversity.

> *Data File:* **NATIONS**
> *Task:* **Mapping**
> *Variable 1:* **56) MULTI-CULT**
> *View:* **Map**

MULTI-CULT -- MULTI-CULTURALISM:ODDS THAT ANY 2 PERSONS WILL DIFFER IN THEIR RACE, RELIGION, ETHNICITY (TRIBE),OR LANGUAGE GROUP (STARK)

The lighter the color, the less diversity. Despite several exceptions, overall the least diversity appears in parts of Europe, Asia, and Central America, while the highest diversity appears in Africa.

In the surveys included in the NATIONS data set, respondents were asked four questions about their willingness to live, respectively, near Jews, foreigners, Muslims, and members of another race. The NATIONS data set reports the percent of each nation's respondents who said they would *not* want members of these groups to be their neighbors. These variables indicate the extent of the nations' racial and ethnic prejudice. Let's take a brief moment to map each of these variables. (You'll also find it interesting to rank the results of each map.)

Data File: **NATIONS**
Task: **Mapping**
➤ *Variable 1:* **77) ANTI-SEM.**
➤ *View:* **Map**

ANTI-SEM. -- PERCENT WHO WOULD NOT WANT JEWS AS NEIGHBORS (WVS)

Data File: **NATIONS**
Task: **Mapping**
➤ *Variable 1:* **78) ANTI-FORGN**
➤ *View:* **Map**

ANTI-FORGN -- PERCENT WHO WOULD NOT WANT FOREIGNERS AS NEIGHBORS (WVS, 1997)

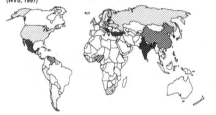

Data File: **NATIONS**
Task: **Mapping**
➤ *Variable 1:* **79) ANTI-MUSLM**
➤ *View:* **Map**

ANTI-MUSLM -- PERCENT WHO WOULD NOT WANT MUSLIMS AS NEIGHBORS (WVS, 1997)

Data File: **NATIONS**
Task: **Mapping**
➤ *Variable 1:* **80) RACISM**
➤ *View:* **Map**

RACISM -- PERCENT WHO WOULD NOT WANT MEMBERS OF ANOTHER RACE AS NEIGHBORS (WVS, 1997)

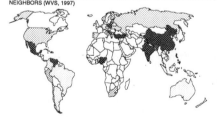

Discovering Sociology

Generally, all four maps look very similar. The most prejudiced nations are generally found in Eastern Europe and parts of Asia.

Why are the people of some nations more likely than those in other nations to be prejudiced? Perhaps education makes a difference. As people acquire a formal education, they tend to learn about other societies and cultures and begin to appreciate the lives of people with different racial and ethnic backgrounds from their own. If this is true, the nations with the highest levels of education should be least prejudiced. Let's examine this possibility with the variable measuring the percent of the population that would not want Jews as neighbors.

> Data File: **NATIONS**
> ➤ Task: **Scatterplot**
> ➤ Dependent Variable: **77) ANTI-SEM.**
> ➤ Independent Variable: **116) EDUCATION**
> ➤ Display: **Reg. Line**

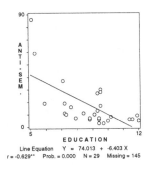

Line Equation Y = 74.013 + -6.403 X
r = -0.629** Prob. = 0.000 N = 29 Missing = 145

The higher a nation's education, the lower its degree of anti-Semitism (r = −.63**).

RACE AND ETHNICITY IN THE UNITED STATES

We now turn our attention to the United States. Let's first get a picture of the U.S. racial and ethnic composition using Census data. As we do, see where your home state fits in.

> ➤ Data File: **STATES**
> ➤ Task: **Mapping**
> ➤ Variable 1: **13) %WHITE**
> ➤ View: **Map**

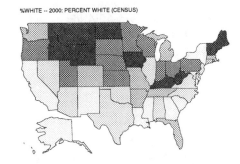

%WHITE -- 2000: PERCENT WHITE (CENSUS)

This map uses U.S. Census data to indicate the percent of each state's population that is white; the darker the color, the higher the percent white. As you can see, this percent tends to be highest in the upper Midwest and in upper New England, and lowest in the South and a few other states, including New Mexico, California, Alaska, and Hawaii.

Data File: **STATES**
Task: **Mapping**
➤ Variable 1: **14) %BLACK**
➤ View: **Map**

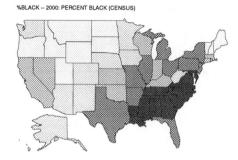

The highest proportion of African Americans live in the Southeast and, more generally, east of the Mississippi River.

Data File: **STATES**
Task: **Mapping**
➤ Variable 1: **15) %AMER.IND.**
➤ View: **Map**

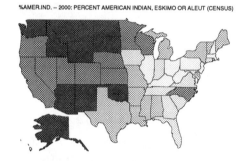

The states that have the highest proportion of Native Americans lie west of the Mississippi River.

Data File: **STATES**
Task: **Mapping**
➤ Variable 1: **16) %ASIAN/PI**
➤ View: **Map**

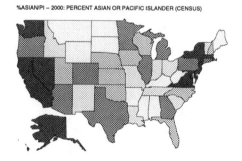

The two major regions for Asians and Pacific Islanders are in the Far West and in the Northeast.

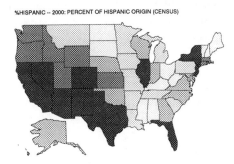

%HISPANIC -- 2000: PERCENT OF HISPANIC ORIGIN (CENSUS)

> Data File: **STATES**
> Task: **Mapping**
➤ Variable 1: **17) %HISPANIC**
➤ View: **Map**

The Southwest and some states east of the Mississippi have the highest proportion of people of Hispanic origin.

We'll now switch to the GSS and examine the extent and predictors of various views on racial issues.

The GSS includes several items measuring such views. One item asked whether there should be laws against marriages between blacks and whites.

➤ Data File: **GSS**
➤ Task: **Univariate**
➤ Primary Variable: **58) INTERMAR.?**
➤ View: **Pie**

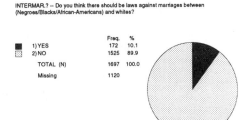

INTERMAR.? -- Do you think there should be laws against marriages between (Negroes/Blacks/African-Americans) and whites?

		Freq.	%
■	1) YES	172	10.1
▨	2) NO	1525	89.9
	TOTAL (N)	1697	100.0
	Missing	1120	

About 10 percent of the U.S. population believes there should be laws prohibiting racial intermarriages. Has this percent declined in the past 30 years?

➤ Data File: **HISTORY**
➤ Task: **Historical Trends**
➤ Variable: **6) INTERMAR.?**

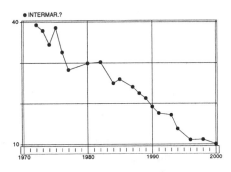

Percent agreeing with laws against racial intermarriage

The percent of the U.S. population favoring laws prohibiting racial intermarriages has declined considerably since the early 1970s.

Our analysis will now focus on white people, and our dependent variable will be another GSS question, "How would you respond to a close relative marrying a black person?" (which you first encountered in the worksheets for Exercise 5). We'll first see how whites responded to this question by using ExplorIt's subset option.

➤ *Data File:* **GSS**
➤ *Task:* **Univariate**
➤ *Primary Variable:* **121) MARRY BLK**
➤ *Subset Variable:* **20) RACE**
➤ *Subset Category:* **Include: 2) White**
➤ *View:* **Pie**

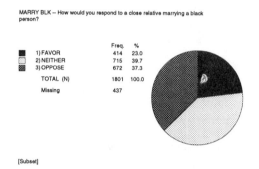

MARRY BLK -- How would you respond to a close relative marrying a black person?

	Freq.	%
1) FAVOR	414	23.0
2) NEITHER	715	39.7
3) OPPOSE	672	37.3
TOTAL (N)	1801	100.0
Missing	437	

[Subset]

The option for selecting a subset variable is located on the same screen you use to select other variables. For this example, select 20) RACE as a subset variable. A window will appear that shows you the categories of the subset variable. Select 2) White as your subset category and choose the [Include] option. Then click [OK] and continue as usual. With this particular subset selected, the results will be limited to the whites in the sample.

About 37 percent of whites would oppose a close relative's marrying an African American.

Historically, the South has been more prejudiced than other U.S. regions. Are white Southerners today more likely than non-Southerners to oppose a relative's marrying an African American?

Data File: **GSS**
➤ *Task:* **Cross-tabulation**
➤ *Row Variable:* **121) MARRY BLK**
➤ *Column Variable:* **17) SOUTH**
➤ *Subset Variable:* **20) RACE**
➤ *Subset Category:* **Include: 2) White**
➤ *View:* **Tables**
➤ *Display:* **Column %**

MARRY BLK by SOUTH

Cramer's V: 0.182 **

		SOUTH		
		SOUTH	NON-SOUTH	TOTAL
MARRY BLK	FAVOR	110	304	414
		18.2%	25.4%	23.0%
	NEITHER	195	520	715
		32.2%	43.5%	39.7%
	OPPOSE	301	371	672
		49.7%	31.0%	37.3%
	Missing	148	289	437
	TOTAL	606	1195	1801
		100.0%	100.0%	

White Southerners are more likely (49.7 percent) than their non-southern counterparts (31.0 percent) to oppose a close relative's marrying an African American (V = .18**).

Let's end this preliminary section with brief tests of two important theories in the study of race relations in the United States. The first is Hubert Blalock's power-threat hypothesis, which assumes that prejudice and hostility toward racial minorities increases during times of economic problems. His theory would lead us to hypothesize that whites whose economic situation has worsened will be more racially prejudiced than whites whose economic circumstances have not worsened. The GSS asked respondents, "During the last few years, has your financial situation been getting better, worse or has

Discovering Sociology

it stayed the same?" Because the "last few years" before this question was put to respondents in 2000 were a time of economic expansion in the United States, even people whose financial situation stayed the same might have felt frustrated over their own lack of economic improvement. Drawing from Blalock's hypothesis, we thus hypothesize that respondents whose financial situation grew worse or only stayed the same will be more racially prejudiced than those whose situation improved. Our dependent variable will be a measure of racial prejudice that asked respondents whether they think blacks' higher poverty rate stems from their "lack of motivation or willpower to pull themselves out of poverty."

> Data File: **GSS**
> Task: **Cross-tabulation**
> ➤ Row Variable: **109) RACE DIF4**
> ➤ Column Variable: **162) CHANGE $?**
> ➤ Subset Variable: **20) RACE**
> ➤ Subset Category: **Include: 2) White**
> ➤ View: **Tables**
> ➤ Display: **Column %**

RACE DIF4   by   CHANGE $?

Cramer's V: 0.124 **

		CHANGE $?				
		BETTER	THE SAME	WORSE	Missing	TOTAL
RACE DIF4	YES	284	261	137	2	682
		44.4%	57.2%	55.7%		50.8%
	NO	356	195	109	2	660
		55.6%	42.8%	44.3%		49.2%
	Missing	385	364	134	9	892
	TOTAL	640	456	246	13	1342
		100.0%	100.0%	100.0%		

Our hypothesis is supported: whites whose financial situation worsened or only remained the same are more racially prejudiced than whites whose situation improved (V = .12**).

The other theory we will briefly test concerns the effects among whites of knowing or living with people of other races. Many sociologists theorize that whites who know people of another race or ethnicity get to see them as real people and not just as stereotypes and thus are less likely to hold prejudicial feelings toward them. However, it is also possible that, as the old saying goes, "familiarity breeds contempt." If this is true, then prejudicial feelings should be stronger among whites who know members of another race or ethnicity than among whites who have no such acquaintanceships. For the moment, let's hypothesize that knowing people of color is associated among whites with reduced prejudice toward that race or ethnicity as a whole. This time we will test our hypothesis with whites' views about Hispanics.

> Data File: **GSS**
> Task: **Cross-tabulation**
> ➤ Row Variable: **123) MARRY HSP**
> ➤ Column Variable: **172) KNW HISP**
> ➤ Subset Variable: **20) RACE**
> ➤ Subset Category: **Include: 2) White**
> ➤ View: **Tables**
> ➤ Display: **Column %**

MARRY HSP   by   KNW HISP

Cramer's V: 0.218 **

		KNW HISP			
		YES	NO	Missing	TOTAL
MARRY HSP	FAVOR	235	53	56	288
		32.0%	22.0%		29.5%
	NEITHER	366	93	40	459
		49.9%	38.6%		47.1%
	OPPOSE	133	95	7	228
		18.1%	39.4%		23.4%
	Missing	12	22	1126	1160
	TOTAL	734	241	1229	975
		100.0%	100.0%		

Our hypothesis is supported. Whites who know any Hispanics are substantially less likely (18.1 percent) than whites who do not know any Hispanics (39.4 percent) to oppose a close relatives marrying someone who is Hispanic (V = .22**). This result suggests that interracial friendships help reduce racial and ethnic prejudice.

WORKSHEET

NAME:

COURSE:

DATE:

REVIEW QUESTIONS

Based on the first part of this exercise, answer True or False to the following items:

Africa tends to have more multicultural diversity than other continents.	T	F
People in South America tend to be more prejudiced than those on other continents.	T	F
The states that have the highest proportion of Native Americans lie east of the Mississippi River.	T	F
About one-third of the U.S. population believes there should be laws prohibiting racial intermarriages.	T	F
Racial prejudice in the United States has declined since the early 1970s.	T	F
Among whites, racial prejudice seems to rise if economic circumstances worsen.	T	F

EXPLORIT QUESTIONS

1. After the tragic events of September 11, 2001, when planes hijacked by terrorists crashed into the World Trade Center, the Pentagon, and a rural area in Pennsylvania, attention around the world focused on Muslims. Reports of physical and verbal attacks against Muslims surfaced in the United States and several other nations. We saw in the preliminary section of this exercise that the proportion of people who would not want Muslims as neighbors varies around the world. The GSS does not include such a direct measure of prejudice against Muslims, but did ask respondents whether they thought Muslims who have settled in the United States have made an important contribution to this country, some contribution, or little contribution. Keep in mind that this question was asked about a year before the events of September 11.

   a. If you had to guess, what percent of the GSS will say that Muslims have made an important contribution? _____%

      ➤ *Data File:* **GSS**
         ➤ *Task:* **Univariate**
      ➤ *Primary Variable:* **147) MUSLM CONT**
            ➤ *View:* **Pie**

   b. What percent of the GSS actually said that Muslims have made an important contribution? _____%

c. If this question were asked of a nationwide sample today, do you think this percent would increase, decrease, or stay about the same? Explain your answer

2. Some scholars say that people who are prejudiced against a particular racial or ethnic group also tend to be prejudiced against other such groups. We can test this hypothesis by seeing whether whites who would not want a close relative marrying an African American are also less likely than other whites to say that Muslims have made an important contribution to this country.

> Data File: **GSS**
> ➤ Task: **Cross-tabulation**
> ➤ Row Variable: **147) MUSLM CONT**
> ➤ Column Variable: **121) MARRY BLK**
> ➤ Subset Variable: **20) RACE**
> ➤ Subset Category: **Include 2) WHITE**
> ➤ View: **Tables**
> ➤ Display: **Column %**

a. Which statement below best describes the results of this cross-tabulation?

1. People who oppose a marriage between a close relative and an African American are especially likely to think that Muslims have made only some or little contribution to the United States.

2. People who oppose a marriage between a close relative and an African American are especially likely to think that Muslims have made an important contribution to the United States.

b. Do these results support the hypothesis that people who are prejudiced against one racial or ethnic group also tend to be prejudiced toward other such groups?   Yes   No

c. Can we assume that opposition to a marriage between a close relative and an African American somehow causes someone also to minimize the contributions that Muslims have made to the United States?

1. Yes, because that is precisely what the results of this cross-tabulation indicate.

2. No, because both of these attitudes might simply reflect underlying prejudicial views that someone may hold.

3. Muslims are one of the many immigrant groups in the United States. Continue your study of immigration by examining the immigration rates (the number of new immigrants admitted per 10,000 population) to each state.

    > *Data File:* **STATES**
    > *Task:* **Mapping**
    > *Variable 1:* **23) IMMIGRANTS**
    > *View:* **List: Rank**

    a. Which state has the highest immigration rate? _____

    b. What is its rate? _____

    c. What is the rate of the state in which your college or university is located? _____

    d. Which state has the lowest immigration rate? _____

    e. What is its rate? _____

4. The GSS asks several questions about immigration, which has been a controversial topic during the last decade as many immigrants, both legal and illegal, have come into the United States. One item in the GSS asked, "Do you think the number of immigrants from foreign countries who are permitted to come to the United States to live should be increased or decreased?"

    a. If you had to guess, what percent of the GSS will say that immigration should be decreased? _____%

    > *Data File:* **GSS**
    > *Task:* **Univariate**
    > *Primary Variable:* **126) IMMIGRANTS**
    > *View:* **Pie**

    b. What percent of the GSS actually said that immigration should be decreased? _____%

    c. Do responses to this question show that attitudes toward immigration became more negative during the 1990s?
        1. Yes, because so many people say that immigration should be decreased.
        2. No, because we do not know what percent of the public in 1990 thought that immigration should be decreased.

9. Switching to views about racial policies, the GSS asks whether the government has spent too little, too much, or about the right amount to improve the conditions of African Americans.

> *Data File:* **GSS**
> *Task:* **Cross-tabulation**
> ➤ *Row Variable:* **37) BLACK $**
> ➤ *Column Variable:* **20) RACE**
> ➤ *View:* **Tables**
> ➤ *Display:* **Column %**

a. About one-half of African Americans feel that the government has spent too little
   to improve the conditions of blacks.   T   F

b. Only 1 percent of African Americans feel that the government has spent too much
   to improve the conditions of blacks.   T   F

c. In responding to this question, a greater percent of whites say "too much" than say
   either "about right" or "too little."   T   F

d. Do the results of this table support conflict theory's view that social groups
   hold beliefs according to their vested interests in maintaining and improving
   their socioeconomic position in society?   Yes   No

10. In the preliminary section of this exercise, we saw that anti-Semitism was lower in more educated nations. Let's see whether we find a similar relationship when we consider the percent who do not want people of other races as neighbors.

> ➤ *Data File:* **NATIONS**
> ➤ *Task:* **Scatterplot**
> ➤ *Dependent Variable:* **80) RACISM**
> ➤ *Independent Variable:* **116) EDUCATION**
> ➤ *View:* **Reg. Line**

a. What is the value of r?   r = _____

b. Is opposition to living near members of other races lower in the more
   educated nations?   Yes   No

c. What would be a sociological explanation that helps account for the answer you just gave?

♦ EXERCISE 8 ♦
GENDER AND GENDER INEQUALITY

Tasks: Mapping, Scatterplot, Univariate, Historical Trends, Cross-tabulation
Data Files: NATIONS, STATES, GSS, HISTORY

The different roles expected of females and males are our *genders*. Whereas sex is a biological category determined at the moment of conception, gender is a social and cultural one. Therefore, gender and gender roles differ from one society to another. Even within a given society, various people have different expectations of how girls and boys and women and men should think and behave.

If, according to the sociological perspective, our social backgrounds influence our behavior, attitudes, and life chances, then gender illustrates the sociological perspective perhaps more than any other social category. Simply put, gender has a profound influence on many aspects of how we think, of how we act, and of our life chances. *Gender socialization* refers to gender's influence on how we think and act, while *gender inequality* refers to gender's influence on our life chances. In speaking of gender inequality, sociologists emphasize that men are the dominant sex and women the subordinate sex. When it comes to such things as wealth, power, prestige, and other life outcomes, men often fare much better than women. To acknowledge such gender inequality is not to suggest there has to be such inequality. Rather, an understanding of the sources and manifestations of gender inequality is crucial if we are to achieve a society in which women and men are equal.

The data sets included with this workbook have many variables related to gender, far too many for us to use all of them. We'll look at just a few, and begin by trying to get some idea of international differences in gender inequality and in views on some important gender issues. Then we'll turn to the United States to see what difference being female or male makes in behavior, attitudes, and life chances. The worksheets will give you the opportunity to explore some gender topics on your own.

CROSS-CULTURAL DIFFERENCES IN GENDER AND GENDER INEQUALITY

People in nations around the world differ in their views about gender issues and in their degree of gender equality. Your NATIONS data set includes a variable listing the percentage of each country's population who agree that "what women really want is a home and children." Whether or not you agree with this belief, it obviously represents a traditional view of women's gender role. Let's see how responses to this question differ across the world.

► *Data File:* **NATIONS**
► *Task:* **Mapping**
► *Variable 1:* **52) HOME&KIDS**
► *View:* **Map**

HOME&KIDS -- PERCENT WHO AGREE THAT WHAT "WOMEN REALLY WANT IS A HOME AND CHILDREN" (WVS)

The darker the color, the higher the percent who feel that women want a home and children most of all. The nations that have the highest percent of residents who feel this way tend to be in Eastern Europe and the underdeveloped world, while the nations whose residents take a more contemporary view—those lighter in color—tend to be in Western Europe and North America.

Data File: **NATIONS**
Task: **Mapping**
Variable 1: **52) HOME&KIDS**
► *View:* **List: Rank**

RANK	CASE NAME	VALUE
1	Lithuania	97.00
2	India	94.00
3	Slovak Republic	93.00
3	Czech Republic	93.00
5	Latvia	90.00
5	Bulgaria	90.00
5	Russia	90.00
8	Turkey	88.00
8	Nigeria	88.00
10	Estonia	85.00

The variation in responses to this question is remarkable. At the high end, 97 percent of Lithuanians and 94 percent of the people of India think that women want a home and children most of all. At the low end, only 25 percent of Denmark's residents feel this way. Note that although the United States is near the low end on this question, more than half of U.S. respondents, or 56 percent, still think that women want a home and children above all. Canada stands 13 percent lower, at 43 percent. Thus the United States is a bit more traditional than Canada on this one aspect of women's gender role.

Now let's look at international variation in gender inequality. Although there are many ways to measure this concept, we'll use a variable that lists the average female years of schooling as a percentage of the male years of schooling. Thus, the lower the percentage, the less education women have compared to men.

<table>
<tr><td align="right">Data File:</td><td>**NATIONS**</td></tr>
<tr><td align="right">Task:</td><td>**Mapping**</td></tr>
<tr><td align="right">➤ Variable 1:</td><td>**47) M/F EDUC.**</td></tr>
<tr><td align="right">➤ View:</td><td>**Map**</td></tr>
</table>

M/F EDUC. -- AVERAGE FEMALE YEARS OF SCHOOLING AS A PERCENTAGE OF AVERAGE MALE YEARS (CALCULATED)

The lighter the color, the less education women have compared to men. Women's education lags farthest behind men's in the underdeveloped world, primarily in the nations of Africa and parts of Asia.

Do beliefs about gender roles vary with gender equality? It makes sense to think that the most traditional beliefs are found in the nations with the least gender equality. Let's see whether this is true by using our previous attitudinal measure on women's role in the home along with an overall measure of gender equality that takes into account several aspects of women's lives.

<table>
<tr><td align="right">Data File:</td><td>**NATIONS**</td></tr>
<tr><td align="right">➤ Task:</td><td>**Scatterplot**</td></tr>
<tr><td align="right">➤ Dependent Variable:</td><td>**52) HOME&KIDS**</td></tr>
<tr><td align="right">➤ Independent Variable:</td><td>**48) FEM POWER**</td></tr>
<tr><td align="right">➤ View:</td><td>**Reg. Line**</td></tr>
</table>

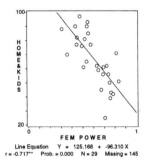

Line Equation   Y = 125.166 + -96.310 X
r = -0.717**   Prob. = 0.000   N = 29   Missing = 145

We see the relationship we expected. People in the nations with the least gender equality are more likely to believe that women want a home and children above all. The correlation is high, $-.72$**.

GENDER AND GENDER INEQUALITY IN THE UNITED STATES

We'll begin our U.S. focus on gender by looking at the STATES data set. We know that, biologically, there's a 50 percent chance that a baby will be a girl and a 50 percent chance that it will be a boy. Thus about half of all babies are female and half are male. However, since men on average die sooner than women, a little more than 50 percent of all people are female. This percent varies slightly by state because of migration patterns. Let's look at the map for the percent of each state that is female.

> ➤ *Data File:* **STATES**
> ➤ *Task:* **Mapping**
> ➤ *Variable 1:* **12) %FEMALE**
> ➤ *View:* **Map**

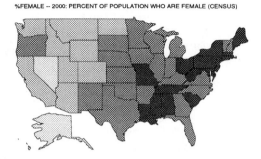

%FEMALE -- 2000: PERCENT OF POPULATION WHO ARE FEMALE (CENSUS)

The states east of the Mississippi River have the highest percentages of female residents, and the states in the West have the lowest percentages. What might explain this geographic pattern?

An important issue to many women and men is abortion. The STATES data set includes a measure of the abortion rate (number of abortions per 1,000 women ages 15–44) for each state.

> *Data File:* **STATES**
> *Task:* **Mapping**
> ➤ *Variable 1:* **40) ABORTIONS**
> ➤ *View:* **Map**

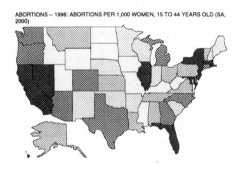

ABORTIONS -- 1996: ABORTIONS PER 1,000 WOMEN, 15 TO 44 YEARS OLD (SA, 2000)

No clear geographic pattern emerges here, although the states on both the east and the west coasts do seem to have the highest abortion rates.

> *Data File:* **STATES**
> *Task:* **Mapping**
> *Variable 1:* **40) ABORTIONS**
> ➤ *View:* **List: Rank**

RANK	CASE NAME	VALUE
1	Nevada	44.6
2	New York	41.1
3	New Jersey	35.8
4	California	33.0
5	Florida	32.0
6	Massachusetts	29.3
7	Hawaii	27.3
8	Maryland	26.3
9	Illinois	26.1
10	Rhode Island	24.4

Nevada and New York have the highest abortion rates, while Wyoming has the lowest.

Let's turn to the GSS and begin with some views on women's roles. We'll first examine the percent of respondents who agree that "it is much better for everyone involved if the man is the achiever outside the home and the woman takes care of the home and family." Do you agree or disagree with this statement?

Discovering Sociology

➤ *Data File:* **GSS**
➤ *Task:* **Univariate**
➤ *Primary Variable:* **105) WIFE@HOME**
➤ *View:* **Pie**

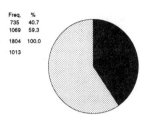

WIFE@HOME – AGREE OR DISAGREE?: It is much better for everyone involved if the man is the achiever outside the home and the woman takes care of the home and family.

		Freq.	%
■	1) AGREE	735	40.7
▨	2) DISAGREE	1069	59.3
	TOTAL (N)	1804	100.0
	Missing	1013	

About 41 percent of the U.S. population agrees that it is much better if men are the achievers outside the home and women take care of the home and family.

This question touches on women's traditional home and family role and indicates that a significant minority of the public—more than 40 percent—continues to accept this traditional role. Has this acceptance declined in the last 30 years? Let's look at the trend for responses to this item.

➤ *Data File:* **HISTORY**
➤ *Task:* **Historical Trends**
➤ *Variable:* **35) WIFE@HOME**

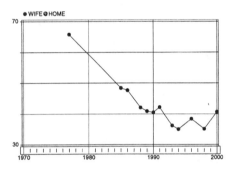

Percent saying women should take care of running their homes

The belief that women should stay at home has declined considerably during the last 30 years, but it has risen slightly since the early 1990s.

One of the most important consequences of gender for life chances lies in the workplace, where women's earnings continue to lag far behind men's. To illustrate this, let's look at the annual earnings—combined into three categories—of women and men who work full-time.

➤ *Data File:* **GSS**
➤ *Task:* **Cross-tabulation**
➤ *Row Variable:* **27) OWN INCOME**
➤ *Column Variable:* **19) GENDER**
➤ *Subset Variable:* **1) FULL TIME?**
➤ *Subset Category:* **Include: 1) Yes**
➤ *View:* **Tables**
➤ *Display:* **Column %**

OWN INCOME   by   GENDER

Cramer's V: 0.268 **

		GENDER		
		FEMALE	MALE	TOTAL
OWN INCOME	$0K-17.4K	181	95	276
		28.6%	13.3%	20.5%
	17.5K-34.9	278	246	524
		44.0%	34.5%	39.0%
	$35K +	173	372	545
		27.4%	52.2%	40.5%
	Missing	86	93	179
	TOTAL	632	713	1345
		100.0%	100.0%	

The option for selecting a subset variable is located on the same screen you use to select other variables. For this example, select 1) FULL TIME? as a subset variable. A window will appear that shows you the categories of the subset variable. Select 1) Yes as your subset category and choose the [Include] option. Then click [OK] and continue as usual.

Among women, 28.6 percent indicate that their personal income is in the lowest income bracket. Among men, this percentage is only 13.3 percent. Conversely, men are more likely (52.2 percent) than women (27.4 percent) to be in the highest bracket (V = .27**). But does this difference persist when we look only at people who have college degrees? Today most college students, male and female, want to get a degree to increase their future earnings. Let's see whether gender affects the income students can expect to receive after graduating from college. We'll have to use two subset variables this time.

Data File: **GSS**
Task: **Cross-tabulation**
Row Variable: **27) OWN INCOME**
Column Variable: **19) GENDER**
Subset Variable 1: **1) FULL TIME?**
Subset Category: **Include: 1) Yes**
➤ *Subset Variable 2:* **15) EDUCATION**
➤ *Subset Category:* **Include: 4) Coll. Grad**
➤ *View:* **Tables**
➤ *Display:* **Column %**

OWN INCOME   by   GENDER

Cramer's V: 0.223 **

		GENDER		
		FEMALE	MALE	TOTAL
OWN INCOME	$0K-17.4K	28	13	41
		14.4%	5.7%	9.7%
	17.5K-34.9	68	53	121
		35.1%	23.0%	28.5%
	$35K +	98	164	262
		50.5%	71.3%	61.8%
	Missing	21	29	50
	TOTAL	194	230	424
		100.0%	100.0%	

To add a subset variable, return to the same screen you use to select other variables. For this example, select 15) EDUCATION as a second subset variable. A window will appear that shows you the categories of the subset variable. Select 4) Coll. Grad as your subset category and choose the [Include] option. Both 15) EDUCATION and 1) FULL TIME? are now listed as subset variables. Then click [OK] and continue as usual.

Even though a greater percent of both sexes with B.A. degrees have high incomes than was true in the previous example for the whole sample, women continue to lag behind men. They are more likely to be in the lowest income bracket (about a 9-percentage-point difference) and less likely to be in the highest bracket (about a 21-percentage-point difference). The relationship is strong and statistically significant (V = .22**). Why do you think this gender gap exists?

Some scholars think this gender gap exists partly because women are "slowed down" more than men by the responsibilities of parenthood. Will a gender gap in income exist if we rule out this factor by limiting our analysis to full-time workers who have never had any children? Let's find out.

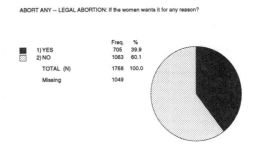

Data File: **GSS**
Task: **Cross-tabulation**
Row Variable: **27) OWN INCOME**
Column Variable: **19) GENDER**
Subset Variable 1: **1) FULL TIME?**
Subset Category: **Include: 1) Yes**
Subset Variable 2: **15) EDUCATION**
Subset Category: **Include: 4) Coll. Grad**
➤ Subset Variable 2: **12) # CHILDREN**
➤ Subset Category: **Include: 0) NONE**
➤ View: **Tables**
➤ Display: **Column %**

OWN INCOME   by   GENDER

Cramer's V: 0.223 **

		GENDER		
		FEMALE	MALE	TOTAL
OWN INCOME	$0K-17.4K	17	8	25
		18.7%	7.5%	12.6%
	17.5K-34.9	37	34	71
		40.7%	31.8%	35.9%
	$35K +	37	65	102
		40.7%	60.7%	51.5%
	Missing	12	13	25
	TOTAL	91	107	198
		100.0%	100.0%	

A gender gap in income exists even for full-time, college-educated people who have never had any children. Women with these backgrounds are about 11 percent more likely than their male counterparts to be in the lowest income bracket and 20 percent less likely to be in the highest bracket (V = .22**). Since we have now ruled out the effects of parenthood, how would you explain this wide gender gap in income for people with college degrees who work full-time?

Finally, let's turn to one of the most controversial gender-related issues today, abortion. The GSS asks whether a woman should be allowed to have a legal abortion if she wants one "for any reason."

Data File: **GSS**
➤ Task: **Univariate**
➤ Primary Variable: **90) ABORT ANY**
➤ View: **Pie**

ABORT ANY -- LEGAL ABORTION: If the women wants it for any reason?

		Freq.	%
■	1) YES	705	39.9
▨	2) NO	1063	60.1
	TOTAL (N)	1768	100.0
	Missing	1049	

Just under 40 percent of the U.S. population thinks a woman should be allowed to have a legal abortion for any reason.

How has this changed during the last 30 years?

> *Data File:* **HISTORY**
> > *Task:* **Historical Trends**
> > *Variable:* **25) ABORT ANY**

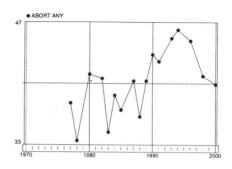

● ABORT ANY

Percent favoring legal abortion for any reason

This figure has risen overall since the 1970s, although it has declined slightly since the early 1990s.

Do you think gender affects support for legalized abortion? It would probably make sense to think that women should support it more than men. Let's find out.

> *Data File:* **GSS**
> > *Task:* **Cross-tabulation**
> *Row Variable:* **90) ABORT ANY**
> *Column Variable:* **19) GENDER**
> > *View:* **Tables**
> > *Display:* **Column %**

ABORT ANY   by   GENDER
Cramer's V: 0.001

		GENDER		
		FEMALE	MALE	TOTAL
ABORT ANY	YES	396	309	705
		39.8%	39.9%	39.9%
	NO	598	465	1063
		60.2%	60.1%	60.1%
	Missing	594	455	1049
	TOTAL	994	774	1768
		100.0%	100.0%	

Surprise! Women are no more likely than men to support legalized abortion. How would you explain this result?

Before proceeding to the worksheets, let's return to the issue of women staying at home. It makes sense to think that people who think preschool children suffer if their mothers work should also be more likely to favor women staying at home.

> *Data File:* **GSS**
> *Task:* **Cross-tabulation**
> *Row Variable:* **105) WIFE@HOME**
> *Column Variable:* **104) PRESCH.WRK**
> > *View:* **Tables**
> > *Display:* **Column %**

WIFE@HOME   by   PRESCH.WRK
Cramer's V: 0.355 **

		PRESCH.WRK			
		AGREE	DISAGREE	Missing	TOTAL
WIFE@HOME	AGREE	490	221	24	711
		58.8%	23.9%		40.5%
	DISAGREE	343	703	23	1046
		41.2%	76.1%		59.5%
	Missing	24	25	964	1013
	TOTAL	833	924	1011	1757
		100.0%	100.0%		

This is a very strong relationship: respondents who feel preschool children suffer if their mothers work are much more likely (58.8 percent) than those who feel these children do not suffer (23.9 percent) to think that women should take care of the home and family (V = .36**).

       Discovering Sociology

Could this relationship be spurious? Recall from Exercise 6 that a relationship is spurious if it disappears once we control for the effects of a third variable that affects both the original independent variable and dependent variable. A variable like gender could render spurious the relationship we just examined under the following reasoning: men may be more likely than women to feel that preschool children suffer if their mothers work *and* that women should take care of the home and family. If so, it is possible that the relationship we found could actually be spurious because we did not control for the effects of gender. We thus need to assess this possibility by examining the original relationship separately for women and men.

Data File: **GSS**
Task: **Cross-tabulation**
Row Variable: **105) WIFE@HOME**
Column Variable: **104) PRESCH.WRK**
➤ Control Variable: **19) GENDER**
➤ View: **Table: Female**
➤ Display: **Column %**

WIFE@HOME   by   PRESCH.WRK
Controls: GENDER: FEMALE
Cramer's V: 0.366 **

		AGREE	DISAGREE	Missing	TOTAL
WIFE@HOME	AGREE	253	144	14	397
		60.8%	24.5%		39.5%
	DISAGREE	163	444	13	607
		39.2%	75.5%		60.5%
	Missing	15	18	524	557
	TOTAL	416	588	551	1004
		100.0%	100.0%		

Data File: **GSS**
Task: **Cross-tabulation**
Row Variable: **105) WIFE@HOME**
Column Variable: **104) PRESCH.WRK**
Control Variable: **19) GENDER**
➤ View: **Table: Male**
➤ Display: **Column %**

WIFE@HOME   by   PRESCH.WRK
Controls: GENDER: MALE
Cramer's V: 0.342 **

		AGREE	DISAGREE	Missing	TOTAL
WIFE@HOME	AGREE	237	77	10	314
		56.8%	22.9%		41.7%
	DISAGREE	180	259	10	439
		43.2%	77.1%		58.3%
	Missing	9	7	440	456
	TOTAL	417	336	460	753
		100.0%	100.0%		

When we control for gender by examining the original relationship separately for women and men, we find that this relationship still exists and continues to be strong for both sexes (V = .37** for women and V = .34** for men). The relationship between the views about preschool children and about women staying at home is thus not spurious when we control for gender, increasing our confidence that the relationship is real.

WORKSHEET

Workbook exercises and software are copyrighted. Copying is prohibited by law.

NAME: _____

COURSE: _____

DATE: _____

EXERCISE

8

REVIEW QUESTIONS

Based on the first part of this exercise, answer True or False to the following items:

College-educated women earn less money than college-educated men.	T F
Western European nations are less likely than Eastern European nations to feel that what a woman really wants is a home and children.	T F
People in nations with the least amount of gender equality are more likely to believe that what women really want is a home and children.	T F
The highest percentage of men is found in states east of the Mississippi River.	T F
The belief that women should stay at home has declined since the 1970s.	T F
Women are more likely than men to support legalized abortion.	T F

EXPLORIT QUESTIONS

1. Does gender affect whether we think that women should take care of running their homes?

> ➤ *Data File:* **GSS**
> ➤ *Task:* **Cross-tabulation**
> ➤ *Row Variable:* **105) WIFE@HOME**
> ➤ *Column Variable:* **19) GENDER**
> ➤ *View:* **Tables**
> ➤ *Display:* **Column %**

a. What percent of men believe that women should take care of running their homes? _____%

b. What percent of women believe that women should take care of running their homes? _____%

c. Is V statistically significant?   Yes   No

d. Keeping in mind statistical significance, are men more likely than women to think that women should take care of running their homes?   Yes   No

    e.   Do the results of this cross-tabulation illustrate the sociological perspective?

         1.   Yes, because they show that gender as an aspect of our social backgrounds influences views on whether women should stay at home.

         2.   No, because they show that gender does not have this influence.

2.   As noted in the preliminary section of this exercise, the view that women should stay at home is a traditional belief. It makes sense to think that people with higher levels of education should be less likely than those with less education to hold this belief.

                Data File: **GSS**
                   Task: **Cross-tabulation**
          Row Variable: **105) WIFE@HOME**
   ➤ Column Variable: **15) EDUCATION**
             ➤ View: **Tables**
         ➤ Display: **Column %**

    a.   Fill in the blank in this sentence: People with college degrees are _____ percent less likely than those without a high school degree to believe that it's better for men to achieve outside the home and for women to run the home and family.

    b.   More than half of all people without a high school degree agree that women should stay at home.   T   F

3.   If education seemingly makes such a difference, should religiosity? Our hypothesis is that people who are more religious should be more likely than those who are less religious to think that men should achieve outside the home and women should stay at home.

                Data File: **GSS**
                   Task: **Cross-tabulation**
          Row Variable: **105) WIFE@HOME**
   ➤ Column Variable: **54) PRAY**
             ➤ View: **Tables**
         ➤ Display: **Column %**

    a.   Now fill in the blank in this sentence: People who pray daily are _____ percent more likely than those who pray less than weekly to believe that it's better for men to achieve outside the home and for women to run the home and family.

    b.   More than half of all people who pray daily agree that women should stay at home.   T   F

4.   Based on the results for Questions 2 and 3, would you also expect that education and religiosity should be related to support for legalized abortion? Obtain two cross-tabulations where 90) ABORT ANY is the dependent (row) variable for each cross-tabulation and 15) EDUCATION is the independent (column) variable for the first cross-tabulation and 54) PRAY is the independent (column) variable for the second cross-tabulation. Remember to percentage your tables.

a. Which statement summarizes the results of the cross-tabulation between education and support for legalized abortion?

    1.   The higher the education, the greater the support for legalized abortion.

    2.   The higher the education, the lower the support for legalized abortion.

    3.   The lower the education, the greater the support for legalized abortion.

b. Which statement summarizes the results of the cross-tabulation between religiosity and support for legalized abortion?

    1.   The more often people pray, the more they support legalized abortion.

    2.   The more often people pray, the less they support legalized abortion.

    3.   The less often people pray, the less they support legalized abortion.

c. Thinking about the results for 105) WIFE@HOME and now for 90) ABORT ANY, do these results suggest that religiosity promotes traditional views and that education promotes nontraditional views?   Yes   No

5.   Let's examine a map showing where abortion is legal throughout the world.

    ➤ *Data File:* **NATIONS**
        ➤ *Task:* **Mapping**
    ➤ *Variable 1:* **14) ABORT LEGL**
        ➤ *View:* **Map**

a. Most of the nations where abortion is legal lie in the Northern Hemisphere.   T   F

b. There is no nation outside of the Northern Hemisphere where abortion is legal.   T   F

c. There is no nation in Africa where abortion is legal.   T   F

d. Abortion is legal in Australia.   T   F

6.   We have seen in the preliminary section of this exercise that women earn lower incomes than men. Let's hypothesize that they should be less happy than men.

    ➤ *Data File:* **GSS**
        ➤ *Task:* **Cross-tabulation**
    ➤ *Row Variable:* **63) HAPPY?**
    ➤ *Column Variable:* **19) GENDER**
        ➤ *View:* **Tables**
        ➤ *Display:* **Column %**

a. Fill in the following line.

	FEMALE	MALE
VERY HAPPY	_____%	_____%

b. Taking into account statistical significance, are men happier than women?   Yes   No

c. Judging from the results of this cross-tabulation, is gender related to happiness?   Yes   No

7. Women tend to be in lower-paying jobs than men and are much more likely than men to be sexually harassed in the workplace. Let's hypothesize that they should be less satisfied than men with their jobs. Our focus is on full-time workers.

> Data File: **GSS**
> Task: **Cross-tabulation**
> ➤ Row Variable: **174) LIKE JOB?**
> ➤ Column Variable: **19) GENDER**
> ➤ Subset Variable: **1) FULL TIME?**
> ➤ Subset Category: **Include: 1) Yes**
> ➤ View: **Tables**
> ➤ Display: **Column %**

a. Is V statistically significant?   Yes   No

b. Keeping in mind the answer you just gave, is our hypothesis supported?   Yes   No

c. Judging from the results of this cross-tabulation, is gender related to job satisfaction?   Yes   No

8. Among women, who should have more exciting lives—women who work full-time or homemakers?

> Data File: **GSS**
> Task: **Cross-tabulation**
> ➤ Row Variable: **66) LIFE**
> ➤ Column Variable: **23) HOMEMAKER?**
> ➤ Subset Variable: **19) GENDER**
> ➤ Subset Category: **Include: 1) Female**
> ➤ View: **Tables**
> ➤ Display: **Column %**

a. What percent of women who work full-time say their lives are exciting?   _____%

b. What percent of women who are homemakers say their lives are exciting?   _____%

c. Is V statistically significant?   Yes   No

d. Judging from the results of this table, are women who work full-time more likely
    than those who are homemakers to think that life is exciting?   Yes   No

e. More than half of the women who work full-time say their lives are exciting.   T   F

9. In the preliminary section of this exercise, we observed international differences in the belief that
    women "really want" a home and children. What might account for this international variation?
    Perhaps this belief will be more common in nations whose people are less educated.

    ➤ *Data File:* **NATIONS**
    ➤ *Task:* **Scatterplot**
    ➤ *Dependent Variable:* **52) HOME&KIDS**
    ➤ *Independent Variable:* **116) EDUCATION**
    ➤ *View:* **Reg. Line**

    a. After looking at the shape of the scatterplot, complete the following sentence: The _____
        a nation's educational level, the _____ its acceptance of the belief that women want a
        home and children above all.

    b. This scatterplot suggests that:
        1. Greater levels of education are related to lower acceptance of a traditional behavior.
        2. Lower levels of education are related to lower acceptance of a traditional behavior.

10. Female circumcision (or, as it's sometimes called, genital mutilation) in Africa is an issue that con-
    cerns many women's rights activists around the world. In many parts of the world, a girl's clitoris and
    other genitalia are cut off before the age of 10. Let's see which countries in Africa practice female cir-
    cumcision most often by using a variable measuring the percent of women who have had all of their
    external sexual organs removed.

    Data File: **NATIONS**
    ➤ *Task:* **Mapping**
    ➤ *Variable 1:* **49) SEX MUTIL**
    ➤ *View:* **List: Rank**

    a. Female circumcision is highest in Egypt.   T   F

    b. Uganda is one of two nations in which female circumcision is lowest.   T   F

    c. Female circumcision is higher in Gambia than in Mali.   T   F

    d. Female circumcision is lower in Kenya than in Niger.   T   F

11. For our last questions in these worksheets, we return to the GSS, which asked respondents how likely it is that "a woman won't get a job or promotion while an equally or less qualified man gets one instead."

    a. Would you say this outcome is likely or not likely?   Likely   Not Likely

    b. What percent of the GSS sample do you think will say this outcome is likely?   _____%

    c. Obtain the appropriate frequency distribution for 128) DISC WOMAN and indicate here the percent of the sample that did, in fact, say that this outcome of job discrimination against women was likely.   _____%

    d. Did more than two-thirds of the sample say that this outcome of job discrimination against women was likely?   Yes   No

12. Earlier in this exercise we saw that women and men held the same opinions on some important issues concerning gender roles. Do you think they will also be equally likely to say that the type of job discrimination just described is likely?

>   *Data File:* **GSS**
> ➤   *Task:* **Cross-tabulation**
> ➤   *Row Variable:* **128) DISC WOMAN**
> ➤ *Column Variable:* **19) GENDER**
> ➤   *View:* **Tables**
> ➤   *Display:* **Column %**

    a. Which statement best describes the results of this table?

        1.   Women are more likely than men to think that this type of job discrimination against women occurs.

        2.   Women are less likely than men to think that this type of job discrimination occurs.

        3.   Taking into account V, there is no gender difference in this view.

    b. Compare these results to those you found when you examined the relationship between gender and attitudes on women's roles at home and on abortion legality. Summarize this comparison.

    c. What might account for the results of the comparison you just made?

13. Select a variable from the GSS data set that you think might be related to support for legalized abortion for any reason and obtain the appropriate cross-tabulation. Make sure you treat the variable you select as the independent (column) variable and that you properly percentage your table.

    a. What variable did you select (name and number)? _____

    b. What relationship do you expect to find?

    c. Print out the percentaged results for your table and turn them in with your assignment.

    d. Is V statistically significant?   Yes   No

    e. In complete sentences, summarize the results of your analysis. Also indicate what conclusion you drew from the table based on the relationship you expected to find.

◆ EXERCISE 9 ◆

AGE AND AGING

Tasks: Mapping, Univariate, Scatterplot, Cross-tabulation, Historical Trends
Data Files: NATIONS, STATES, GSS, HISTORY

As we get older, society's view of us changes. TV shows and commercials feature the young and extol the excitement of youth, while both often portray the elderly as forgetful and frail and even as buffoons. Our society's emphasis on youthfulness leads many older Americans to try to prevent or minimize the physiological effects of aging with things like wrinkle creams and plastic surgery. But time marches on, and eventually many do become frail and see their spouses, friends, and acquaintances develop health problems and, sometimes, die. Many elderly people, especially women whose husbands have already passed away, live by themselves. It's certainly not an easy time of life, but neither is it as lamentable as it's often portrayed in the media. The elderly often lead happy and healthy lives and certainly don't fit the stereotypes that characterize them in the United States and many other nations.

Your workbook's data sets include several variables related to age and aging that we will explore in the next few pages. We'll look at the geographic distribution of older people and at the difference, if any, that age makes in various attitudes, behaviors, and life chances.

AGE AND AGING IN INTERNATIONAL PERSPECTIVE

Life expectancy differs dramatically around the world. Let's see which nations have the longest and shortest expected life spans.

➤ *Data File:* **NATIONS**
➤ *Task:* **Mapping**
➤ *Variable 1:* **18) LIFE EXPCT**
➤ *View:* **Map**

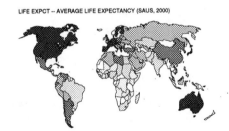

LIFE EXPCT -- AVERAGE LIFE EXPECTANCY (SAUS, 2000)

The lighter the color, the lower the life expectancy. The nations of Africa and much of Asia tend to have the shortest life spans, the nations of North America and Europe the longest.

Euthanasia, or terminating the life of someone with a fatal medical condition, is one of the more controversial issues today. It especially affects the elderly: while the young can have fatal medical conditions, the old are much more likely to suffer from them. Views on euthanasia differ widely around the world. The NATIONS data set includes a variable listing the percent of each nation's residents who believe euthanasia is acceptable.

Data File: **NATIONS**
Task: **Mapping**
➤ Variable 1: **100) EUTHANASIA**
➤ View: **Map**

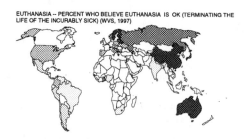

EUTHANASIA -- PERCENT WHO BELIEVE EUTHANASIA IS OK (TERMINATING THE LIFE OF THE INCURABLY SICK) (WVS, 1997)

The pattern in this map isn't as clear as the patterns in other maps we've been seeing, where there have been clear differences between the developed and underdeveloped nations. In particular, the European nations seem to have divergent views on euthanasia; some are very likely to think it's acceptable, while others are much more disapproving.

What explains why some nations are more likely than others to approve of euthanasia? Many variables might come to mind, but let's look at religiosity—how religious people are. Does it makes sense to think that the more religious a nation's people, the less likely they are to approve of euthanasia? Let's find out.

Data File: **NATIONS**
➤ Task: **Scatterplot**
➤ Dependent Variable: **100) EUTHANASIA**
➤ Independent Variable: **72) CH.ATTEND**
➤ View: **Reg. Line**

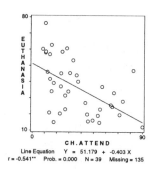

Line Equation   Y = 51.179 + -0.403 X
r = -0.541**   Prob. = 0.000   N = 39   Missing = 135

Our hypothesis is supported: the more frequent the religious attendance, the lower the support for euthanasia. The correlation, r, is a high −.54**.

AGE AND AGING IN THE UNITED STATES

About 13 percent of the U.S. population is 65 and older. Yet this percentage varies across the country. Some states have higher proportions of the elderly, while others have lower proportions.

> *Data File:* **STATES**
> > *Task:* **Mapping**
> *Variable 1:* **9) %OVER 64**
> > *View:* **Map**

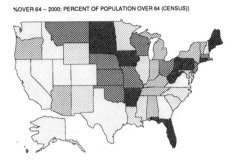
%OVER 64 -- 2000: PERCENT OF POPULATION OVER 64 (CENSUS))

The states with the highest proportions of the elderly tend to be in Florida, several states in the Northeast, and the Midwest. The West generally has the lowest percent of older people. Why do you think this is so?

We can also see what percent of each state's population lives in nursing homes.

> *Data File:* **STATES**
> *Task:* **Mapping**
> *Variable 1:* **82) %NURS.HOME**
> > *View:* **Map**

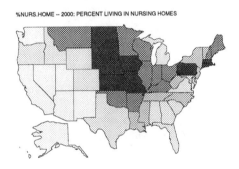
%NURS.HOME -- 2000: PERCENT LIVING IN NURSING HOMES

Not surprisingly, this map looks similar to the one for percent elderly.

> *Data File:* **STATES**
> > *Task:* **Scatterplot**
> *Dependent Variable:* **82) %NURS.HOME**
> *Independent Variable:* **9) %OVER 64**
> > *View:* **Reg. Line**

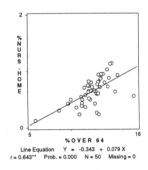
Line Equation   Y = -0.343 + 0.079 X
r = 0.643**   Prob. = 0.000   N = 50   Missing = 0

The scatterplot confirms that the percent who are elderly in a state is strongly related (r = .64**) to the state's percent in nursing homes. One puzzle that arises from inspecting the maps is why Florida's nursing home percent is not in the top tier, given that its percent who are elderly is so high. Are its elderly residents healthier than those elsewhere? Wealthier (and hence able to afford better care than nursing homes)? Are they more likely to have their children living near them? What do you think?

We now turn to the GSS to see how, if at all, older Americans differ from those in younger age groups. We often hear that the elderly, many of whom are on fixed incomes, are especially likely to have financial problems. Let's see whether age is related to family income from all sources.

➤ Data File: **GSS**
➤ Task: **Cross-tabulation**
➤ Row Variable: **26) FAM INCOME**
➤ Column Variable: **14) AGE 65+**
➤ View: **Tables**
➤ Display: **Column %**

FAM INCOME   by   AGE 65+
Cramer's V: 0.206 **

		AGE 65+			
		18-64	65+	Missing	TOTAL
FAM INCOME	$0K-24.9K	665	225	2	890
		32.1%	58.9%		36.3%
	$25K-49.9K	635	87	1	722
		30.7%	22.8%		29.4%
	$50K +	770	70	1	840
		37.2%	18.3%		34.3%
	Missing	256	101	4	361
	TOTAL	2070	382	8	2452
		100.0%	100.0%		

Older Americans are more likely to fall into the low-income category, and less likely to fall into the high-income category (V = .21**). On the whole, they are indeed poorer than their younger counterparts.

Does the elderly's worse financial situation translate into dissatisfaction with their finances?

Data File: **GSS**
Task: **Cross-tabulation**
➤ Row Variable: **85) SAT.$?**
➤ Column Variable: **14) AGE 65+**
➤ View: **Tables**
➤ Display: **Column %**

SAT.$?   by   AGE 65+
Cramer's V: 0.131 **

		AGE 65+			
		18-64	65+	Missing	TOTAL
SAT $?	PRETTY WEL	630	202	2	832
		27.2%	42.3%		29.8%
	MORE/LESS	1063	194	4	1257
		45.9%	40.6%		45.0%
	NOT SATIS.	624	82	2	706
		26.9%	17.2%		25.3%
	Missing	9	5	0	14
	TOTAL	2317	478	8	2795
		100.0%	100.0%		

Far from it! Those over 65 are *more* likely (42.3 percent) to say they're "pretty well" satisfied with their financial situation than are those under 65 (27.2 percent). The elderly are also less likely to say they're "not satisfied" (V = .13**). How would you explain this surprising result?

Other common perceptions of the elderly are that they spend many days by themselves and have numerous health problems. Let's see whether the GSS data confirm this. We'll first see whether the elderly are more likely to live alone.

Data File: **GSS**
Task: **Cross-tabulation**
➤ Row Variable: **22) ALONE?**
➤ Column Variable: **14) AGE 65+**
➤ View: **Tables**
➤ Display: **Column %**

ALONE?   by   AGE 65+
Cramer's V: 0.275 **

		AGE 65+			
		18-64	65+	Missing	TOTAL
ALONE?	ALONE	483	255	4	738
		20.8%	52.8%		26.3%
	NOT ALONE	1843	228	4	2071
		79.2%	47.2%		73.7%
	TOTAL	2326	483	8	2809
		100.0%	100.0%		

The elderly are much more likely (52.8 percent) to live alone than are younger people (20.8 percent) (V = .28**).

Now we will see whether older Americans also report worse health.

Data File: **GSS**
Task: **Cross-tabulation**
➤ Row Variable: **65) HEALTH**
➤ Column Variable: **14) AGE 65+**
➤ View: **Tables**
➤ Display: **Column %**

HEALTH   by   AGE 65+
Cramer's V: 0.235 **

		AGE 65+			
		18-64	65+	Missing	TOTAL
HEALTH	EXCELLENT	635	65	2	700
		33.2%	16.1%		30.2%
	GOOD	934	166	3	1100
		48.8%	41.1%		47.4%
	FAIR/POOR	346	173	0	519
		18.1%	42.8%		22.4%
	Missing	411	79	3	493
	TOTAL	1915	404	8	2319
		100.0%	100.0%		

As expected, older Americans are much more likely than their younger counterparts to say that their health is only fair or poor (V = .24**).

If older people are much more likely to live alone and to have poor health, it makes sense to hypothesize that they should feel less happy than younger people. Let's find out.

Data File: **GSS**
Task: **Cross-tabulation**
➤ Row Variable: **63) HAPPY?**
➤ Column Variable: **14) AGE 65+**
➤ View: **Tables**
➤ Display: **Column %**

HAPPY?   by   AGE 65+
Cramer's V: 0.044

		AGE 65+			
		18-64	65+	Missing	TOTAL
HAPPY?	VERY HAPPY	715	164	2	879
		31.2%	34.5%		31.7%
	PRET.HAPPY	1345	253	5	1598
		58.7%	53.2%		57.7%
	NOT TOO	233	59	1	292
		10.2%	12.4%		10.5%
	Missing	33	7	0	40
	TOTAL	2293	476	8	2769
		100.0%	100.0%		

Surprise! The two age groups report similar levels of happiness. V is not statistically significant. This result certainly doesn't fit the common assumption that the elderly lead despondent lives.

Still another perception of the elderly is that they suffer from senility and other problems having to do with thinking processes. Although the GSS doesn't have any direct measures of these problems, it does include a vocabulary test of 10 words, some of them easy, others difficult. Presumably, if the elderly do suffer from these problems, they should fare much worse on this test than their younger counterparts. Let's look at the number of correct answers for each group.

	Data File:	**GSS**
	Task:	**Cross-tabulation**
➤	Row Variable:	**146) #CRCT.WORD**
➤	Column Variable:	**14) AGE 65+**
➤	View:	**Tables**
➤	Display:	**Column %**

#CRCT.WORD   by   AGE 65+
Cramer's V: 0.077 *

		AGE 65+			
		18-64	65+	Missing	TOTAL
#CRCT.WORD	0-5	399	104	0	503
		36.9%	45.0%		38.3%
	6	251	37	1	288
		23.2%	16.0%		21.9%
	7 OR MORE	432	90	0	522
		39.9%	39.0%		39.8%
	Missing	1244	252	7	1503
	TOTAL	1082	231	8	1313
		100.0%	100.0%		

Older Americans do have lower scores, but the difference is very slight (V = .08*). Moreover, it might be due to their lower educational levels rather than to senility or other such problems. Let's test this assumption by using education as a *control* variable. A control variable is different from a subset variable. As you'll recall, a subset variable allows you to select which survey respondents on a third variable you want to include (or exclude) from your analysis. A control variable, which we'll use here, doesn't simply exclude cases but, instead, breaks down the analysis so that the cases for each category of the control variable are shown as separate tables. In this example, we'll repeat the above analysis of #CRCT.WORD and AGE 65+ but will see separate tables for each category of the EDUCATION variable. If the elderly are not more likely than younger respondents to have lower scores once education is taken into account, then within each education group the elderly should no longer have lower scores.

	Data File:	**GSS**
	Task:	**Cross-tabulation**
	Row Variable:	**146) #CRCT.WORD**
	Column Variable:	**14) AGE 65+**
➤	Control Variable:	**15) EDUCATION**
➤	View:	**Tables (NOT H.S.)**
➤	Display:	**Column %**

#CRCT.WORD   by   AGE 65+
Controls: EDUCATION: NOT H.S.
Cramer's V: 0.175 *

		AGE 65+		
		18-64	65+	TOTAL
#CRCT.WORD	0-5	100	58	158
		63.7%	74.4%	67.2%
	6	34	6	40
		21.7%	7.7%	17.0%
	7 OR MORE	23	14	37
		14.6%	17.9%	15.7%
	Missing	171	81	252
	TOTAL	157	78	235
		100.0%	100.0%	

The option for selecting a control variable is located on the same screen you use to select other variables. For this example, select 15) EDUCATION as a control variable and then click [OK] to continue as usual. Separate tables for each of the 15) EDUCATION categories will now be shown for the 146) #CRCT.WORD and 14) AGE 65+ cross-tabulation.

The above table includes only those respondents who don't have a high school degree. Compare the results for the two age groups before continuing.

➤ *View:* **Tables (H.S. GRAD)**

➤ *Display:* **Column %**

#CRCT.WORD   by   AGE 65+
Controls: EDUCATION: H.S. GRAD
Cramer's V: 0.104

AGE 65+				
#CRCT.WORD	18-64	65+	Missing	TOTAL
0-5	141	30	0	171
	45.6%	42.9%		45.1%
6	79	12	0	91
	25.6%	17.1%		24.0%
7 OR MORE	89	28	0	117
	28.8%	40.0%		30.9%
Missing	360	83	1	444
TOTAL	309	70	1	379
	100.0%	100.0%		

Click the appropriate button at the bottom of the task bar to look at the second (or "next") partial table for 15) EDUCATION.

This table includes those respondents who have only a high school degree. Again, examine the results before continuing.

➤ *View:* **Tables (SOME COLL.)**

➤ *Display:* **Column %**

#CRCT.WORD   by   AGE 65+
Controls: EDUCATION: SOME COLL.
Cramer's V: 0.115

AGE 65+				
#CRCT.WORD	18-64	65+	Missing	TOTAL
0-5	117	9	0	126
	37.9%	23.7%		36.3%
6	82	9	1	91
	26.5%	23.7%		26.2%
7 OR MORE	110	20	0	130
	35.6%	52.6%		37.5%
Missing	395	44	4	443
TOTAL	309	38	5	347
	100.0%	100.0%		

Again, click the appropriate button at the bottom of the task bar to look at the third (or "next") partial table for 15) EDUCATION.

And this table includes those respondents having only some college. Let's look at the final table, college graduates.

➤ *View:* **Tables (COLL. GRAD)**

➤ *Display:* **Column %**

#CRCT.WORD   by   AGE 65+
Controls: EDUCATION: COLL. GRAD
Cramer's V: 0.049

AGE 65+				
#CRCT.WORD	18-64	65+	Missing	TOTAL
0-5	41	5	0	46
	13.4%	11.9%		13.3%
6	55	10	0	65
	18.0%	23.8%		18.7%
7 OR MORE	209	27	0	236
	68.5%	64.3%		68.0%
Missing	313	42	2	357
TOTAL	305	42	2	347
	100.0%	100.0%		

Click the appropriate button at the bottom of the task bar to look at the last (or "next") partial table for 15) EDUCATION.

Our suspicions are confirmed. Once education is taken into account, the elderly's vocabulary scores are not lower than those of their younger counterparts. In fact, they generally tend to be higher.

Now let's turn to views on a few issues that particularly affect the elderly. Earlier we looked at international differences in support for euthanasia. The GSS has a similar question: "When a person has a disease that cannot be cured, do you think doctors should be allowed by law to end the patient's life by some painless means if the patient and his family request it?"

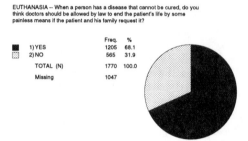

	Data File:	GSS
➤	Task:	Univariate
➤	Primary Variable:	99) EUTHANASIA
➤	View:	Pie

EUTHANASIA -- When a person has a disease that cannot be cured, do you think doctors should be allowed by law to end the patient's life by some painless means if the patient and his family request it?

		Freq.	%
■	1) YES	1205	68.1
▦	2) NO	565	31.9
	TOTAL (N)	1770	100.0
	Missing	1047	

About 68 percent of the sample responded Yes to this question.

➤	Data File:	HISTORY
➤	Task:	Historical Trends
➤	Variable:	12) EUTHANASIA

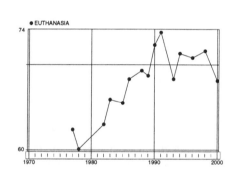

Percent approving of euthanasia

This figure has generally risen since the 1970s.

Let's see whether age affects views on euthanasia. Do you think the elderly should be more in favor or less in favor?

➤	Data File:	GSS
➤	Task:	Cross-tabulation
➤	Row Variable:	99) EUTHANASIA
➤	Column Variable:	14) AGE 65+
➤	View:	Tables
➤	Display:	Column %

EUTHANASIA by AGE 65+
Cramer's V: 0.095 **

		AGE 65+			
		18-64	65+	Missing	TOTAL
EUTHANASIA	YES	1037	166	2	1203
		70.1%	58.0%		68.1%
	NO	443	120	2	563
		29.9%	42.0%		31.9%
	Missing	846	197	4	1047
	TOTAL	1480	286	8	1766
		100.0%	100.0%		

Those over 65 are *less* likely (58 percent) than their younger counterparts (70.1 percent) to approve of euthanasia (V = .10**).

In the NATIONS data set we saw that religiosity, as measured by religious attendance, was negatively related to approval of euthanasia. Will we find a similar relationship in the GSS? We'll use a measure of religious attendance as our independent or column variable.

Data File:	**GSS**				
Task:	**Cross-tabulation**				
Row Variable:	**99) EUTHANASIA**				
➤ Column Variable:	**52) ATTEND**				
➤ View:	**Tables**				
➤ Display:	**Column %**				

EUTHANASIA by ATTEND
Cramer's V: 0.279 **

		ATTEND				
		NEVER	MONTH/YRLY	WEEKLY	Missing	TOTAL
EUTHANASIA	YES	316	620	245	24	1181
		82.7%	73.1%	48.7%		68.1%
	NO	66	228	258	13	552
		17.3%	26.9%	51.3%		31.9%
	Missing	201	492	311	43	1047
	TOTAL	382	848	503	80	1733
		100.0%	100.0%	100.0%		

The higher the religiosity, the lower the support for euthanasia (V = .28**).

As our population ages, adult children of aging parents are increasingly having to make important decisions about where their parents should live. The GSS asks, "As you know, many older people share a home with their grown children. Do you think this is generally a good idea or a bad idea?"

Do you think older people will be more likely than younger ones to think it's a good idea?

Data File:	**GSS**
Task:	**Cross-tabulation**
➤ Row Variable:	**81) LIVE W KID**
➤ Column Variable:	**14) AGE 65+**
➤ View:	**Tables**
➤ Display:	**Column %**

LIVE W KID by AGE 65+
Cramer's V: 0.266 **

		AGE 65+			
		18-64	65+	Missing	TOTAL
LIVE W KID	GOOD IDEA	878	80	4	958
		67.5%	32.3%		61.9%
	BAD IDEA	422	168	2	590
		32.5%	67.7%		38.1%
	Missing	1026	235	2	1263
	TOTAL	1300	248	8	1548
		100.0%	100.0%		

People 65 and older are much *less* likely than those under 65 to think that it's a good idea for older people to share a home with their grown children (V = .27**). Did you expect this result?

WORKSHEET

NAME:

COURSE:

DATE:

EXERCISE
9

REVIEW QUESTIONS

Based on the first part of this exercise, answer True or False to the following items:

The chances of living into your 70s do not generally depend on the region of the world in which you live.	T	F
The more religious a nation's people, the more they approve of euthanasia.	T	F
In the United States, people are more likely to live in a nursing home on the West Coast than in other regions of the country.	T	F
Older Americans are generally less happy than younger Americans.	T	F
Older Americans are less likely than younger Americans to approve of euthanasia.	T	F
Older Americans are more alone than younger Americans.	T	F

Based on the analyses in this exercise so far, would you say that older Americans have worse lives than younger people? Why or why not?

EXPLORIT QUESTIONS

Many people think that older people hold more conservative views than younger people on controversial social issues having to do with morality, partly because they grew up during a time when views were more conservative than they are now. Let's explore possible age differences on several of these issues.

1. We'll first consider sex education in the schools.

> ➤ *Data File:* **GSS**
> ➤ *Task:* **Cross-tabulation**
> ➤ *Row Variable:* **180) SEX ED?**
> ➤ *Column Variable:* **14) AGE 65+**
> ➤ *View:* **Tables**
> ➤ *Display:* **Column %**

Exercise 9: Age and Aging

a. What percent of older people are against sex education in the public schools? _____%

b. What percent of younger people are against sex education in the public schools? _____%

c. Is V statistically significant?   Yes   No

d. Do the elderly hold more conservative views on sex education?   Yes   No

2. Now we'll examine possible age differences in views on divorce.

> *Data File:* **GSS**
> *Task:* **Cross-tabulation**
> ➤ *Row Variable:* **93) DIV.EASY?**
> ➤ *Column Variable:* **14) AGE 65+**
> ➤ *View:* **Tables**
> ➤ *Display:* **Column %**

a. What percent of older people feel that divorce should be made more difficult? _____%

b. What percent of younger people feel that divorce should be made more difficult? _____%

c. Is V statistically significant?   Yes   No

d. Do the elderly hold more conservative views on divorce?   Yes   No

3. Next we'll look at views on premarital sex.

> *Data File:* **GSS**
> *Task:* **Cross-tabulation**
> ➤ *Row Variable:* **94) PREM.SEX**
> ➤ *Column Variable:* **14) AGE 65+**
> ➤ *View:* **Tables**
> ➤ *Display:* **Column %**

a. What percent of older people feel that premarital sex is always wrong? _____%

b. What percent of younger people feel that premarital sex is always wrong? _____%

c. Is V statistically significant?   Yes   No

d. Do the elderly hold more conservative views on premarital sex?   Yes   No

4. Now we'll consider extramarital sex.

> Data File: **GSS**
> Task: **Cross-tabulation**
> ➤ Row Variable: **96) XMAR.SEX**
> ➤ Column Variable: **14) AGE 65+**
> ➤ View: **Tables**
> ➤ Display: **Column %**

a. What percent of older people feel that extramarital sex is always wrong? _____%

b. What percent of younger people feel that extramarital sex is always wrong? _____%

c. Is V statistically significant? Yes No

d. Do the elderly hold more conservative views on extramarital sex? Yes No

5. Are there age differences in views on homosexuality?

> Data File: **GSS**
> Task: **Cross-tabulation**
> ➤ Row Variable: **97) HOMO.SEX**
> ➤ Column Variable: **14) AGE 65+**
> ➤ View: **Tables**
> ➤ Display: **Column %**

a. What percent of older people feel that homosexual sex is always wrong? _____%

b. What percent of younger people feel that homosexual sex is always wrong? _____%

c. Is V statistically significant? Yes No

d. Do the elderly hold more conservative views on homosexual sex? Yes No

6. Finally, we'll examine views on the legalization of marijuana.

> Data File: **GSS**
> Task: **Cross-tabulation**
> ➤ Row Variable: **50) GRASS?**
> ➤ Column Variable: **14) AGE 65+**
> ➤ View: **Tables**
> ➤ Display: **Column %**

a. What percent of older people oppose the legalization of marijuana? _____%

    b. What percent of younger people oppose the legalization of marijuana?   _____%

    c. Is V statistically significant?   Yes   No

    d. Do the elderly hold more conservative views on marijuana?   Yes   No

7.   Based on the analyses in these worksheets, do you think the elderly generally hold more conservative views on social and moral issues than younger people? Why or why not? What sociological explanation accounts for the conclusion you drew from your analysis?

8.   In the beginning of this exercise, we saw that life expectancy varies greatly around the world. What accounts for this variation? An obvious possibility is the wealth or poverty of a nation. Let's see whether the annual national product per capita, the variable we used to measure global stratification in Exercise 6, is related to life expectancy. We'll hypothesize that wealthier nations should have longer life expectancies than poorer nations.

> ➤ *Data File:* **NATIONS**
> ➤ *Task:* **Scatterplot**
> ➤ *Dependent Variable:* **18) LIFE EXPCT**
> ➤ *Independent Variable:* **30) GDP/CAP**
> ➤ *View:* **Reg. Line**

    a. What is the value of r?   r = _____

    b. Is this a weak, moderate, or strong relationship? (Circle one.)   Weak

                                                      Moderate

                                                      Strong

    c. If wealthier nations do have longer life expectancies (or, to turn it around, if poorer nations have shorter life expectancies), what, in your opinion, are the one or two most important reasons for these differences?

9. Perhaps you just mentioned better nutrition as one reason wealthier nations have longer life expectancies. If so, a scatterplot should reveal a positive correlation between better nutrition and longer life expectancy. The NATIONS data set includes a measure of the daily available calories per person (based on a nation's food supply), a rough measure of the nutritional intake in each nation. Let's see whether the nations with higher caloric intake have longer life expectancies.

> Data File: **NATIONS**
> Task: **Scatterplot**
> Dependent Variable: **18) LIFE EXPCT**
> ➤ Independent Variable: **25) CALORIES**
> ➤ View: **Reg. Line**

a. Which statement below best summarizes the results of this scatterplot?

1. Nations with a lower caloric intake have longer life expectancies.
2. Nations with a higher caloric intake have shorter life expectancies.
3. Nations with a higher caloric intake have longer life expectancies.

b. To find out where the United States appears in the scatterplot, click on [Find Case], then select United States from the window that appears, then click [OK]. The dot for the United States will then appear in the scatterplot. Where do you find it?

Upper left

Upper right

Lower left

Lower right

10. Because the elderly may lack transportation options or have reduced physical mobility, they may find it more difficult than younger people to engage in certain leisure-time activities. Let's see whether this is true for social activities. We'll use a GSS variable that asked respondents how often health problems have interfered with their social activities during the past month.

> ➤ Data File: **GSS**
> ➤ Task: **Cross-tabulation**
> ➤ Row Variable: **154) SOC ACTS**
> ➤ Column Variable: **14) AGE 65+**
> ➤ View: **Tables**
> ➤ Display: **Column %**

a. Taking into account V, are people age 65 and older more likely to say that health problems have interfered with their social activity?

Yes   No

b. Did this relationship surprise you? Why or why not?

11. Several of the examples in the preliminary section and worksheets of this exercise address various beliefs about the elderly. Write a brief essay in which you use these examples to discuss the extent to which our beliefs about the elderly turn out not to be true when tested with actual data. (Be sure to provide specific statistical evidence to support your answer.)

◆ EXERCISE 10 ◆

THE FAMILY

Tasks: Mapping, Scatterplot, Univariate, Cross-tabulation, Historical Trends
Data Files: NATIONS, STATES, GSS, HISTORY

Perhaps our most important social institution is the family. Certainly it is the one with which we've had the most contact. Most of us are born into a family and are raised by it until we near the end of our teenage years. Most people later get married and have children or at least live with someone without getting married. The family provides not only food, clothing, shelter, and other necessities, but also socialization, that essential building block of society. As a primary group, the family also provides emotional support for its members.

For better or worse, however, the family has been in a state of flux. Divorce rates began rising dramatically in the 1960s and have leveled off only recently. Single-parent households, either as the result of divorce or as the result of births outside of wedlock, began to increase at about the same time. So did cohabitation, or living together in a romantic relationship without being married. All of these trends have been the subject of considerable, often heated, debate that shows no signs of diminishing.

This exercise examines some key aspects of the family in contemporary life. We'll first look at international variation in some views about the family, and then examine state-by-state variations in divorce and some other family dimensions. We'll finally turn to the GSS to examine opinions on important family issues.

INTERNATIONAL VIEWS ON THE FAMILY

As you might expect, the nations of the world differ widely in their views on various aspects of marriage and the family. Let's first examine the percent who agree that "marriage is an outdated institution."

> ➤ *Data File:* **NATIONS**
> ➤ *Task:* **Mapping**
> ➤ *Variable 1:* **53) WED PASSE'**
> ➤ *View:* **Map**

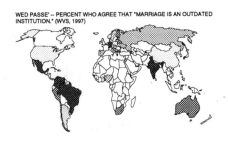

WED PASSE' -- PERCENT WHO AGREE THAT "MARRIAGE IS AN OUTDATED INSTITUTION." (WVS, 1997)

Mexico, India, and several countries in South America and Europe appear to be the most likely to agree that marriage is outdated.

Data File:	**NATIONS**	
Task:	**Mapping**	
Variable 1:	**53) WED PASSE'**	
➤ *View:*	**List: Rank**	

RANK	CASE NAME	VALUE
1	Venezuela	30.7
2	Germany	30.0
3	Brazil	29.5
4	Moldova	29.3
5	Mexico	25.8
5	Switzerland	25.8
7	Colombia	25.3
8	Slovenia	24.9
8	India	24.9
10	Latvia	21.3

Although well under half of all the nations' residents feel marriage is outdated, Venezuela's, with 31 percent, are most likely to feel this way. The United States ranks near the bottom; only 10.5 percent of its residents think marriage is outdated. China, with 8 percent, is least likely to view marriage as outdated. What would explain these differences?

Many of today's controversial issues about the family concern sex and sexuality. One variable in the NATIONS data set measures the percent who think it is "never" acceptable for married people to have an affair (adultery).

Data File:	**NATIONS**
Task:	**Mapping**
➤ *Variable 1:*	**93) EX-MARITAL**
➤ *View:*	**Map**

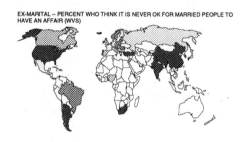

EX-MARITAL -- PERCENT WHO THINK IT IS NEVER OK FOR MARRIED PEOPLE TO HAVE AN AFFAIR (WVS)

No clear geographic pattern really emerges here.

Despite the lack of a geographic pattern, we can still try to understand why the people of some nations are more likely than others to oppose adultery. In previous exercises we have seen that religiosity is related to acceptance of traditional beliefs: the greater the religiosity, the lower the acceptance of traditional beliefs. We can thus hypothesize that higher levels of religiosity in nations around the world predict greater opposition to adultery.

Data File: **NATIONS**
➤ *Task:* **Scatterplot**
➤ *Dependent Variable:* **93) EX-MARITAL**
➤ *Independent Variable:* **72) CH.ATTEND**
➤ *View:* **Reg. Line**

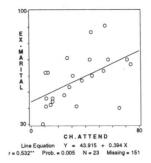

Line Equation   Y = 43.915 + 0.394 X
r = 0.532**   Prob. = 0.005   N = 23   Missing = 151

Our hypothesis is clearly supported: the greater the religiosity, the greater the percent who say it is never okay for married people to have an affair (r = −.53**).

Another controversial sexual issue today is homosexuality. The NATIONS data set includes a measure of the percent who believe homosexuality is "never" acceptable.

Data File: **NATIONS**
➤ *Task:* **Mapping**
➤ *Variable 1:* **95) GAY SEX**
➤ *View:* **Map**

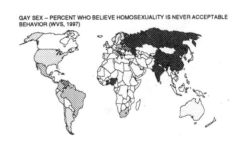

GAY SEX -- PERCENT WHO BELIEVE HOMOSEXUALITY IS NEVER ACCEPTABLE BEHAVIOR (WVS, 1997)

Disapproval of homosexuality is lowest in the nations of Western Europe.

One reason for this international variation in views about homosexuality might be differences in education. In previous exercises we have seen that higher levels of education predict lower acceptance of traditional beliefs. If this is true, then as people become more educated, they become less disapproving of homosexuality. If so, the most educated nations would be less likely than the less-educated nations to say homosexuality is never acceptable. Let's see whether this is so.

Data File: **NATIONS**
➤ *Task:* **Scatterplot**
➤ *Dependent Variable:* **95) GAY SEX**
➤ *Independent Variable:* **116) EDUCATION**
➤ *View:* **Reg. Line**

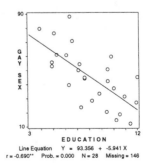

Line Equation   Y = 93.356 + -5.941 X
r = -0.690**   Prob. = 0.000   N = 28   Missing = 146

Our hypothesis is supported: the more educated a nation's residents, the less likely they are to say that homosexuality is never acceptable (r = −.69**).

MARRIAGE AND THE FAMILY IN THE UNITED STATES

Let's take a look at which states have the highest marriage rates. We'll use a measure of marriages per 1,000 population.

➤ *Data File:* **STATES**
    ➤ *Task:* **Mapping**
➤ *Variable 1:* **24) MARRIAGES**
    ➤ *View:* **Map**

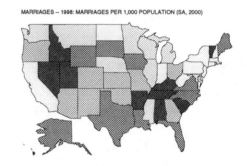

MARRIAGES -- 1998: MARRIAGES PER 1,000 POPULATION (SA, 2000)

No clear geographic pattern emerges here, because the states with the highest marriage rates are in different parts of the country.

    Data File: **STATES**
      Task: **Mapping**
  Variable 1: **24) MARRIAGES**
    ➤ *View:* **List: Rank**

RANK	CASE NAME	VALUE
1	Nevada	79.5
2	Hawaii	17.5
3	Arkansas	15.1
4	Tennessee	14.8
5	Idaho	12.4
6	Alabama	11.4
7	Kentucky	11.2
8	South Carolina	10.7
9	Utah	10.1
10	Vermont	9.9

Nevada's marriage rate soars far above that of any other state. The popular image of people getting married abruptly in Reno might not be too far off the mark!

We can also see which states have the highest proportion of divorced people.

Discovering Sociology

Data File: **STATES**
Task: **Mapping**
➤ Variable 1: **25) DIVORCES**
➤ View: **Map**

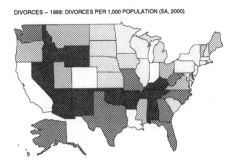
DIVORCES -- 1998: DIVORCES PER 1,000 POPULATION (SA, 2000)

Divorce is most common west of the Mississippi. Why do you think that's so? Is there something about living out west that helps lead to divorce? Do divorced people tend to move west? Or is something else going on?

We turn now to the GSS and begin by continuing to look at divorce. We'll first see what percent of the sample has ever been divorced or separated. (Note that people who have *never* been married are excluded from this calculation.)

➤ Data File: **GSS**
➤ Task: **Univariate**
➤ Primary Variable: **7) DIVORCED?**
➤ View: **Pie**

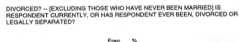
DIVORCED? -- [EXCLUDING THOSE WHO HAVE NEVER BEEN MARRIED] IS RESPONDENT CURRENTLY, OR HAS RESPONDENT EVER BEEN, DIVORCED OR LEGALLY SEPARATED?

	Freq.	%
1) YES	910	43.7
2) NO	1172	56.3
TOTAL (N)	2082	100.0
Missing	735	

About 44 percent of those who have ever been married have been divorced/separated (hereafter referred to as just "divorced").

What predicts your chances of getting divorced? We often hear that your chances are greater if your own parents were divorced. Let's see whether GSS data support this hypothesis.

Data File: **GSS**
➤ Task: **Cross-tabulation**
➤ Row Variable: **7) DIVORCED?**
➤ Column Variable: **21) PARS. DIV?**
➤ View: **Tables**
➤ Display: **Column %**

DIVORCED? by PARS. DIV?
Cramer's V: 0.124 **

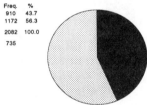

		PARS. DIV?			
		TOGETHER	DIVORCED	Missing	TOTAL
DIVORCED?	YES	581	132	197	713
		39.1%	56.9%		41.5%
	NO	906	100	166	1006
		60.9%	43.1%		58.5%
	Missing	449	166	120	735
	TOTAL	1487	232	483	1719
		100.0%	100.0%		

People whose parents were divorced are about 18 percent more likely to get divorced themselves (V = .12**).

What else predicts divorce? Let's see whether the less religious individuals in the GSS are more likely to be divorced.

Data File: **GSS**
Task: **Cross-tabulation**
Row Variable: **7) DIVORCED?**
➤ Column Variable: **53) HOW RELIG?**
➤ View: **Tables**
➤ Display: **Column %**

DIVORCED?   by   HOW RELIG?
Cramer's V: 0.139 **

		HOW RELIG?			
		STRONG	NOT VERY	Missing	TOTAL
D I V O R C E D ?	YES	265	340	305	605
		35.7%	49.4%		42.3%
	NO	477	348	347	825
		64.3%	50.6%		57.7%
	Missing	196	269	270	735
	TOTAL	742	688	922	1430
		100.0%	100.0%		

People who say they are not very religious are more likely (49.4 percent) than those who say they are strongly religious (35.7 percent) to be divorced (V = .14**). Note that we can't necessarily conclude that religiosity affects divorce, because people might become less religious after getting divorced. The relationship might also be spurious if some third factor, say income, affects both religiosity and the chances for divorce.

Now let's just compare people in terms of their *current* marital status. Which group do you think will be the happiest: those who are married, widowed, divorced (or separated), or single? Let's find out.

Data File: **GSS**
Task: **Cross-tabulation**
➤ Row Variable: **63) HAPPY?**
➤ Column Variable: **6) MARITAL**
➤ View: **Tables**
➤ Display: **Column %**

HAPPY?   by   MARITAL
Cramer's V: 0.185 **

		MARITAL					
		MARRIED	WIDOWED	DIV./SEP.	NEV.MARR	Missing	TOTAL
H A P P Y ?	VERY HAPPY	555	66	110	150	0	881
		43.6%	24.5%	20.3%	21.6%		31.7%
	PRET.HAPPY	650	161	341	450	1	1602
		51.1%	59.9%	62.9%	64.9%		57.7%
	NOT TOO	67	42	91	93	0	293
		5.3%	15.6%	16.8%	13.4%		10.6%
	Missing	6	4	11	19	0	40
	TOTAL	1272	269	542	693	1	2776
		100.0%	100.0%	100.0%	100.0%		

Marital status is certainly related to happiness (V = .19**); married people are the most likely to say they're very happy.

The GSS asks married people, "Taking things all together, how would you describe your marriage? Would you say that your marriage is very happy, pretty happy, or not too happy?"

Data File: **GSS**
➤ Task: **Univariate**
➤ Primary Variable: **64) HAP.MARR.?**
➤ View: **Pie**

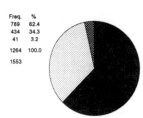

HAP.MARR.? -- IF CURRENTLY MARRIED: Taking things all together, how would you describe your marriage? Would you say that your marriage is very happy, pretty happy, or not too happy?

		Freq.	%
■	1) VERY HAPPY	789	62.4
▨	2) PRET.HAPPY	434	34.3
▩	3) NOT TOO	41	3.2
	TOTAL (N)	1264	100.0
	Missing	1553	

About 62 percent of married people describe their marriage as very happy.

➤ Data File: **HISTORY**
➤ Task: **Historical Trends**
➤ Variable: **14) HAP.MARR**

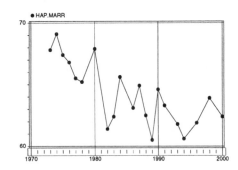

Percent saying their marriage is very happy

This percent remains largely unchanged for the past 30 years with only a 5 percent decline.

Given the problems that low income can cause for families, it makes sense to think that poorer marriages will be less happy. Does family income predict marital happiness?

➤ Data File: **GSS**
➤ Task: **Cross-tabulation**
➤ Row Variable: **64) HAP.MARR.?**
➤ Column Variable: **26) FAM INCOME**
➤ View: **Tables**
➤ Display: **Column %**

HAP.MARR.?   by   FAM INCOME
Cramer's V: 0.072 *

		FAM INCOME				
		$0K-24.9K	$25K-49.9K	$50K +	Missing	TOTAL
HAP.MARR.?	VERY HAPPY	109	201	391	88	701
		57.1%	60.0%	66.2%		62.8%
	PRET.HAPPY	72	119	189	54	380
		37.7%	35.5%	32.0%		34.0%
	NOT TOO	10	15	11	5	36
		5.2%	4.5%	1.9%		3.2%
	Missing	701	388	250	214	1553
	TOTAL	191	335	591	361	1117
		100.0%	100.0%	100.0%		

The social class differences here are statistically significant (V = .07*) but rather small. Did you expect a stronger relationship?

Many scholars think that people who are more religious have happier marriages. Our hypothesis is that people who pray frequently will report happier marriages than those who pray less often.

Data File: **GSS**

Task: **Cross-tabulation**

Row Variable: **64) HAP.MARR.?**

➤ Column Variable: **54) PRAY**

➤ View: **Tables**

➤ Display: **Column %**

HAP.MARR.?   by   PRAY
Cramer's V: 0.090 *

		PRAY				
		DAILY	WEEKLY	< WEEKLY	Missing	TOTAL
HAP.MARR.?	VERY HAPPY	245	81	74	389	400
		67.5%	61.4%	53.2%		63.1%
	PRET.HAPPY	112	47	59	216	218
		30.9%	35.6%	42.4%		34.4%
	NOT TOO	6	4	6	25	16
		1.7%	3.0%	4.3%		2.5%
	Missing	415	154	182	802	1553
	TOTAL	363	132	139	1432	634
		100.0%	100.0%	100.0%		

People who pray daily are more likely (67.5 percent) to report their marriages are very happy than those who pray less than weekly (53.2 percent). The results are statistically significant (V = .09**).

Could this relationship between religiosity and marital happiness be spurious? Suppose that people with low levels of education are more likely than those with higher education both to be more religious and, for different reasons, to have less happy marriages. Controlling for education could thus render spurious the original relationship between religiosity and marital happiness. Let's find out.

Data File: **GSS**

Task: **Cross-tabulation**

Row Variable: **64) HAP.MARR.?**

Column Variable: **54) PRAY**

➤ Control Variable: **15) EDUCATION**

➤ View: **Tables**

➤ Display: **Column %**

HAP.MARR.?   by   PRAY
Controls: EDUCATION: NOT H.S
Cramer's V: 0.137
Warning: Potential significance problem. Check row and column totals.

		PRAY				
		DAILY	WEEKLY	< WEEKLY	Missing	TOTAL
HAP.MARR.?	VERY HAPPY	24	8	9	51	41
		53.3%	61.5%	42.9%		51.9%
	PRET.HAPPY	19	5	12	33	36
		42.2%	38.5%	57.1%		45.6%
	NOT TOO	2	0	0	7	2
		4.4%	0.0%	0.0%		2.5%
	Missing	96	21	23	177	317
	TOTAL	45	13	21	268	79
		100.0%	100.0%	100.0%		

When we examine the original relationship between praying and marital happiness separately for each of the education categories, we discover that this relationship is no longer statistically significant for any of the education groups. The relationship between praying and marital happiness does, then, turn out to be spurious once we control for education, and we cannot conclude (pending much more sophisticated analysis beyond the scope of this workbook) that religiosity promotes marital happiness.

In the NATIONS data set, we examined attitudes on adultery. Let's see what respondents say in the GSS.

Data File:	**GSS**
➤ *Task:*	**Univariate**
➤ *Primary Variable:*	**96) XMAR.SEX**
➤ *View:*	**Pie**

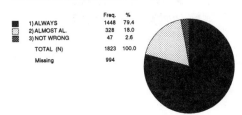

XMAR.SEX -- What is your opinion about a married person having sexual relations with someone other than the marriage partner -- is it always wrong, almost always wrong, wrong only sometimes, or not wrong at all?

		Freq.	%
■	1) ALWAYS	1448	79.4
▦	2) ALMOST AL.	328	18.0
▦	3) NOT WRONG	47	2.6
	TOTAL (N)	1823	100.0
	Missing	994	

About 79 percent of the U.S. public thinks adultery is always wrong.

➤ *Data File:*	**HISTORY**
➤ *Task:*	**Historical Trends**
➤ *Variable:*	**15) XMAR.SEX**

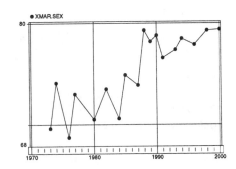

Percent saying extramarital sex is always wrong

This percent has generally risen during the last 30 years.

Now let's look at views on premarital sex. Traditionally, sex was considered acceptable only if it occurred between spouses. Many people still feel that way, but many others think sex outside of marriage is acceptable. The GSS asked what respondents think about a man and a woman having "sex relations before marriage."

➤ *Data File:*	**GSS**
➤ *Task:*	**Univariate**
➤ *Primary Variable:*	**94) PREM.SEX**
➤ *View:*	**Pie**

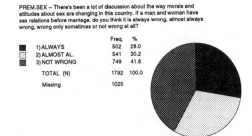

PREM.SEX -- There's been a lot of discussion about the way morals and attitudes about sex are changing in this country. If a man and woman have sex relations before marriage, do you think it is always wrong, almost always wrong, wrong only sometimes or not wrong at all?

		Freq.	%
■	1) ALWAYS	502	28.0
▦	2) ALMOST AL.	541	30.2
▦	3) NOT WRONG	749	41.8
	TOTAL (N)	1792	100.0
	Missing	1025	

Exactly 28 percent think premarital sex is always wrong, about 30 percent think it is almost always wrong, and almost 42 percent think it is not wrong.

> *Data File:* **HISTORY**
> > *Task:* **Historical Trends**
> > *Variable:* **16) PREM.SEX**

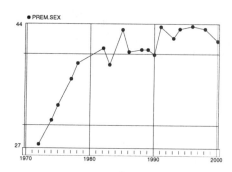

Percent saying premarital sex is not wrong

The figure for thinking that premarital sex is *not* wrong has risen since 30 years ago, but has recently declined slightly.

Do you think we'll find a gender difference on views about premarital sex?

> *Data File:* **GSS**
> > *Task:* **Cross-tabulation**
> *Row Variable:* **94) PREM.SEX**
> *Column Variable:* **19) GENDER**
> > *View:* **Tables**
> > *Display:* **Column %**

PREM.SEX   by   GENDER
Cramer's V: 0.078 **

		GENDER		
		FEMALE	MALE	TOTAL
PREM.SEX	ALWAYS	309	193	502
		30.4%	24.9%	28.0%
	ALMOST AL.	315	226	541
		31.0%	29.2%	30.2%
	NOT WRONG	393	356	749
		38.6%	45.9%	41.8%
	Missing	571	454	1025
	TOTAL	1017	775	1792
		100.0%	100.0%	

Women are slightly more likely than men to oppose premarital sex (V = .08**). Why do you think this difference exists?

What about religiosity? Since we have previously seen that religiosity predicts greater acceptance of traditional beliefs, we will hypothesize that the more religious people are, the more likely they are to think premarital sex is wrong.

Data File: **GSS**
Task: **Cross-tabulation**
Row Variable: **94) PREM.SEX**
> *Column Variable:* **53) HOW RELIG?**
> *View:* **Tables**
> *Display:* **Column %**

PREM.SEX   by   HOW RELIG?
Cramer's V: 0.324 **

		HOW RELIG?			
		STRONG	NOT VERY	Missing	TOTAL
PREM.SEX	ALWAYS	278	103	121	381
		46.3%	16.9%		31.5%
	ALMOST AL.	163	212	166	375
		27.2%	34.9%		31.0%
	NOT WRONG	159	293	297	452
		26.5%	48.2%		37.4%
	Missing	338	349	338	1025
	TOTAL	600	608	922	1208
		100.0%	100.0%		

Discovering Sociology

Religiosity is strongly related to views on premarital sex, because people who are strongly religious are more likely (46.3 percent) than those who are not very religious (16.9 percent) to think premarital sex is always wrong (V = .32**).

WORKSHEET

NAME:

COURSE:

DATE:

EXERCISE

10

REVIEW QUESTIONS

Based on the first part of this exercise, answer True or False to the following items:

In several nations, more than half the population thinks marriage is an outdated institution.	T	F
The more educated a nation's people, the more they disapprove of homosexuality.	T	F
In the United States, divorce is most common in the Northeast.	T	F
In the United States, disapproval of adultery has declined steadily since the early 1970s.	T	F
Men are more likely than women to oppose premarital sex.	T	F
Married people are happier than never-married people.	T	F

EXPLORIT QUESTIONS

1. An important family issue today revolves around discipline of children, specifically spanking. Many people accept the motto "Spare the rod and spoil the child." Others think that spanking should not be used. The GSS asks whether respondents agree or disagree that "it is sometimes necessary to discipline a child with a good, hard spanking."

    > ➤ *Data File:* **GSS**
    > ➤ *Task:* **Univariate**
    > ➤ *Primary Variable:* **98) SPANKING**
    > ➤ *View:* **Pie**

    What percent of the population agrees that spanking is sometimes necessary? _____%

2. Would you expect a gender difference in approval of spanking?

    > *Data File:* **GSS**
    > ➤ *Task:* **Cross-tabulation**
    > ➤ *Row Variable:* **98) SPANKING**
    > ➤ *Column Variable:* **19) GENDER**
    > ➤ *View:* **Tables**
    > ➤ *Display:* **Column %**

Exercise 10: The Family

157

a.   What percent of women approve of spanking?   _____%

b.   What percent of men approve of spanking?   _____%

c.   Is V statistically significant?   Yes   No

d.   Are men more likely than women to approve of spanking?   Yes   No

e.   The results of this cross-tabulation suggest that gender is related to attitudes about spanking.   T   F

3.   The motto "Spare the rod and spoil the child" comes from the Bible. Our hypothesis is that people who think the Bible is the actual word of God should be more likely to approve of spanking.

> Data File: **GSS**
> Task: **Cross-tabulation**
> Row Variable: **98) SPANKING**
> ➤ Column Variable: **57) BIBLE**
> ➤ View: **Tables**
> ➤ Display: **Column %**

a.   Complete this sentence: People who think the Bible is the actual word of God are _____ percent more likely than those who think the Bible is an ancient book of fables to agree that spanking is sometimes necessary.

b.   Taking into account V, do these results support the hypothesis?   Yes   No

4.   Now select another independent variable that you think might predict approval for spanking. Obtain the appropriate cross-tabulation, and then answer these questions.

a.   Which variable did you select (number and name)?   _____

b.   How and why did you think this variable would predict approval for spanking?

c.   Briefly describe the results you obtained in your cross-tabulation.

    d. Looking at all the results in these worksheets, discuss in a brief essay the factors that appear to increase support among Americans for spanking.

5. The GSS asks respondents who report being romantically involved whether they live with their spouse or partner. By using a subset for people who have never been married, we can determine what percentage of unmarried people who are romantically involved live with their partner.

> *Data File:* **GSS**
> ➤ *Task:* **Univariate**
> ➤ *Primary Variable:* **131) TOGETHER**
> ➤ *Subset Variable:* **6) MARITAL**
> ➤ *Subset Category:* **Include: 4) Nev. Marr.**
> ➤ *View:* **Pie**

    a. Fewer than half of the people who are romantically involved actually live with their partner.   T   F

    b. Is this percentage higher or lower than what you would have expected?   Higher   Lower

6. Many people think they will be happier if they move in with their partners. Our hypothesis is that people who live with their partner are happier than those who do not live with their partner.

> *Data File:* **GSS**
> ➤ *Task:* **Cross-tabulation**
> ➤ *Row Variable:* **63) HAPPY?**
> ➤ *Column Variable:* **131) TOGETHER**
> ➤ *Subset Variable:* **6) MARITAL**
> ➤ *Subset Category:* **Include: 4) Nev. Marr.**
> ➤ *View:* **Tables**
> ➤ *Display:* **Column %**

    a. Complete the following sentence: People who live with their partner are _____ percent more likely than those who do not live with their partner to say they are very happy.

    b. Judging from the results of this table, including V, is the hypothesis supported?   Yes   No

7. Let's see whether the divorce rate is lower in states with higher rates of church membership.

    ➤ *Data File:* **STATES**
    ➤ *Task:* **Scatterplot**
    ➤ *Dependent Variable :* **25) DIVORCES**
    ➤ *Independent Variable:* **83) CH.MEMBERS**
    ➤ *View:* **Reg. Line**

    a. Complete the following sentence: The higher the church membership rate of a state, the _____ its divorce rate.

    b. Use the [Find Case] option to locate Nevada in the scatterplot.
        In which part of the scatterplot is Nevada located? (Circle one.)

        Lower right

        Upper right

        Lower left

        Upper left

8. The NATIONS data set has a measure of the percent of each nation who think the ideal number of children in a family is three or more.

    ➤ *Data File:* **NATIONS**
    ➤ *Task:* **Scatterplot**
    ➤ *Dependent Variable:* **9) LARGE FAML**
    ➤ *Independent Variable:* **116) EDUCATION**
    ➤ *View:* **Reg. Line**

    a. What is the value of r?   r = _____

    b. Is r statistically significant?   Yes   No

    c. At the nation level, is education linked to preferences for having three or more children in a family?   Yes   No

    d. Use the [Find Case] option to locate the United States in the scatterplot.
        In which part of the scatterplot is the U.S. located? (Circle one.)

        Lower right

        Upper right

        Lower left

        Upper left

9. In the preliminary section of this exercise, we examined international differences in the percent who say that marriage is an outdated practice. Select a variable from the NATIONS data set that you think might be related to the percent who feel this way. Obtain a scatterplot where 53) WED PASSÉ is the dependent variable and the variable you chose is the independent variable. Then answer the following questions.

    a. What independent variable did you select (name and number)? _____

    b. Why did you feel this variable would be related to 53) WED PASSE'? And did you expect a positive or a negative relationship?

    c. Did the results in your scatterplot support your hypothesis? Why or why not?

10. In the preliminary section you mapped a NATIONS measure of the percent who say homosexuality is never acceptable, and you then saw that this percent was lower for nations with greater levels of education. Should a similar relationship exist in the GSS? Our hypothesis is that the percent of GSS respondents who say that homosexual sex is always wrong will decline as their educational levels increase.

    > *Data File:* **GSS**
    > *Task:* **Cross-tabulation**
    > *Row Variable:* **97) HOMO.SEX**
    > *Column Variable:* **15) EDUCATION**
    > *View:* **Tables**
    > *Display:* **Column %**

    a. Complete the following sentence: People without a high school degree are _____ percent more likely than those with a college degree to say that homosexual sex is always wrong.

    b. Taking into account V, do the results of this cross-tabulation support the hypothesis?   Yes   No

11. In the STATES data set we saw that divorce rates differ across the country. Let's see whether the poorer states tend to have higher divorce rates.

> ➤ *Data File:* **STATES**
> ➤ *Task:* **Scatterplot**
> ➤ *Dependent Variable:* **25) DIVORCES**
> ➤ *Independent Variable:* **45) %POOR**
> ➤ *View:* **Reg. Line**

   a. What is the value of r?   r = _____

   b. Does the scatterplot depict a positive relationship or a negative relationship?

                                               Positive   Negative

   c. Use the [Find Case] option to locate Nevada in the scatterplot. In which part of the scatterplot is Nevada located? (Circle one.)

                                               Lower right

                                               Upper right

                                               Lower left

                                               Upper left

   d. Did this result surprise you? Why or why not?

12. In the preliminary section we saw that opposition to extramarital sex, or adultery, has generally risen in the United States during the last 25 years. Let's look at two factors that might affect our views on this controversial topic. The first is gender.

   a. First it's hypothesis time: Do you think women will be more opposed than men to adultery, that men will be more opposed than women, or that there will be no gender difference? (Circle one.)

                                               Women more opposed

                                               Men more opposed

                                               No difference

   b. Obtain a cross-tabulation from the GSS where 96) XMAR.SEX is the dependent (row) variable and 19) GENDER is the independent (column) variable. Percentage your table properly and then answer the following questions:

     What percent of women say extramarital sex is always wrong?   _____%

     What percent of men say extramarital sex is always wrong?   _____%

Is V statistically significant?   Yes   No

Is your hypothesis supported or not supported?   Supported   Not Supported

c. What would be a sociological reason that helps explain the results in your table?

13. Now let's carry out the same exercise for education.

a. We have seen earlier in the workbook that higher levels of education often predict more support for nontraditional behaviors. If this is true, should high school dropouts be more opposed than college graduates to adultery or should they be less opposed?

   1. High school dropouts should be more opposed than college graduates to adultery.

   2. High school dropouts should be less opposed than college graduates to adultery.

b. Obtain a cross-tabulation from the GSS where 96) XMAR.SEX is the dependent (row) variable and 15) EDUCATION is the independent variable (column) variable. Is the hypothesis supported?   Yes   No

c. What would be a sociological reason that helps explain the results in your table?

◆ EXERCISE 11 ◆

EDUCATION AND RELIGION

Tasks: Mapping, Scatterplot, Cross-tabulation
Data Files: NATIONS, STATES, GSS

As social institutions, education and religion play a very important role in most people's lives. First, they both function as agents of socialization. Through schooling and religious training we learn many of our values and beliefs. As we've seen in some previous exercises, education is often associated with acceptance of nontraditional beliefs and practices, whereas religiosity is often associated with just the opposite.

Second, both education and religion help to integrate our society. Not only do we learn common values from both institutions, we also interact with others as we go to school and attend religious services. As Emile Durkheim emphasized long ago, such social interaction is important for social stability.

Third, both education and religion help determine our life chances. More educated people typically have greater opportunities for high-paying jobs and the benefits that high incomes often bring. Unfortunately, the opposite is true as well: lack of education often dooms people to low incomes, poor health, and other negative life chances. In the United States and elsewhere, religious affiliation has historically been associated with one's placement on the socioeconomic ladder. In the United States, for example, certain Protestant denominations have enjoyed high socioeconomic status and political power, while other Protestant denominations have ranked lower on both scores. Although times have changed, historically U.S. Catholics and Jews were victims of discrimination in the schools and the workplace and also victims of violence directed at them because of their religion. Hate crimes against Jews continue to occur.

Education and religion around the world exhibit similar characteristics. The level of a nation's education is critical for its people's life chances and often for their beliefs. Religious affiliation and practice have a similar international impact. To understand the world today, it's important to appreciate the difference that education and religion make.

This exercise explores some key aspects of education and religion across the world and within the United States. We will primarily focus on what difference they make for life chances and beliefs. We'll first look at education and then at religion and, within each institution's discussion, follow our usual practice of starting with the NATIONS data set and then turning to our two U.S. data sets.

EDUCATION

International Differences in Education and Its Correlates

Let's first map education to see which parts of the world are most and least educated.

> *Data File:* **NATIONS**
> *Task:* **Mapping**
> *Variable 1:* **116) EDUCATION**
> *View:* **Map**

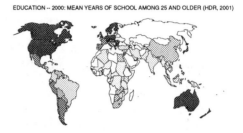

EDUCATION -- 2000: MEAN YEARS OF SCHOOL AMONG 25 AND OLDER (HDR, 2001)

The countries of North America and Western Europe clearly have the highest average education and Africa generally has the lowest.

As noted above, we've already seen that international differences in education are often associated with differences in the acceptance of nontraditional beliefs and practices. Let's see whether education is associated with one additional belief, abortion. The NATIONS data set includes a measure of the percent who approve of an abortion when the mother's health is at risk.

> *Data File:* **NATIONS**
> *Task:* **Scatterplot**
> *Dependent Variable:* **15) MOM HEALTH**
> *Independent Variable:* **116) EDUCATION**
> *View:* **Reg. Line**

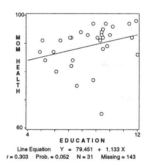

Line Equation   Y = 79.451 + 1.133 X
r = 0.303   Prob. = 0.052   N = 31   Missing = 143

The more educated a nation's residents are, the more likely they are to approve of abortion when the mother's health is in danger (r = .30). Click on [Outlier] and you'll find that Ireland's support for an abortion for this reason is much lower than would be predicted from its relatively high level of education, perhaps because so many of its residents are Catholic. With Ireland removed from the analysis, the relationship between education and support for abortion when the mother's health is at risk becomes much stronger (r = .44**).

Education in the United States

Has it ever occurred to you that people in some regions of the United States are more educated than those in other regions? Let's take a look. Our first map will be the percent of a state's 25-and-over population who have at least a college degree, and the second map will be the percent who never completed high school.

➤ *Data File:* **STATES**
    ➤ *Task:* **Mapping**
➤ *Variable 1:* **51) COLLEGE**
    ➤ *View:* **Map**

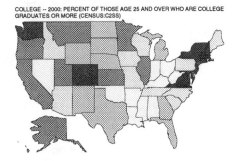

College graduates are most likely to be found in several states around the country, with the greatest concentration of these states on the East Coast. Click on [List:Rank] and you'll see that the percent of adults 25 and older with a college degree ranges from Massachusetts at the top with 35 percent to West Virginia at the bottom with only 14 percent.

Data File: **STATES**
    Task: **Mapping**
➤ *Variable 1:* **52) HS DROPOUT**
    ➤ *View:* **Map**

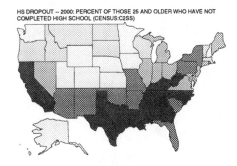

The southern states have the highest proportion of high school dropouts. It makes sense that the least educated states should also be the poorest.

Data File: **STATES**
    ➤ *Task:* **Scatterplot**
➤ *Dependent Variable:* **45) %POOR**
➤ *Independent Variable:* **52) HS DROPOUT**
    ➤ *View:* **Reg. Line**

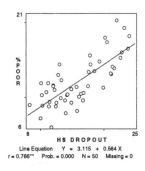

The higher the proportion of high school dropouts, the poorer the state (r = .77**). Note that we can't necessarily conclude that low education produces high poverty, because high poverty may lead to lower education.

Let's turn now to the GSS. The "states attainment" literature emphasizes the impact that our parents' educational level has on our socioeconomic status. To illustrate this influence, we will first examine the relationship between parents' education and the education you eventually acquire. Our independent variable has three categories: both parents lacked a high school degree, both parents had a high school degree, and both parents had at least a college degree. For simplicity's sake, respondents whose parents had different degrees are excluded.

➤ *Data File:* **GSS**
➤ *Task:* **Cross-tabulation**
➤ *Row Variable:* **15) EDUCATION**
➤ *Column Variable:* **163) PARS.DEGR.**
➤ *View:* **Tables**
➤ *Display:* **Column %**

EDUCATION by PARS.DEGR.
Cramer's V: 0.380 **

		PARS.DEGR.				
		BOTH <HS	BOTH HS	BOTH COLL.	Missing	TOTAL
EDUCATION	NOT H.S.	168	28	0	291	196
		32.2%	5.2%	0.0%		16.1%
	H.S. GRAD	191	176	12	444	379
		36.6%	32.5%	8.0%		31.2%
	SOME COLL.	104	193	37	457	334
		19.9%	35.6%	24.7%		27.5%
	COLL. GRAD	59	145	101	399	305
		11.3%	26.8%	67.3%		25.1%
	Missing	2	1	1	8	12
	TOTAL	522	542	150	1599	1214
		100.0%	100.0%	100.0%		

Your parents' education is very strongly related to your own education (V = .38**). While only 11.3 percent of the respondents with parents who dropped out of high school attained college degrees, 67.3 percent of the respondents with college-educated parents did so. Based on these figures, we are six times more likely to get a college degree if our parents went to college than if they didn't graduate from high school.

Education is important for many other life chances. One of these is our health.

Data File: **GSS**
Task: **Cross-tabulation**
➤ *Row Variable:* **65) HEALTH**
➤ *Column Variable:* **15) EDUCATION**
➤ *View:* **Tables**
➤ *Display:* **Column %**

HEALTH by EDUCATION
Cramer's V: 0.219 **

		EDUCATION					
		NOT H.S.	H.S. GRAD	SOME COLL.	COLL. GRAD	Missing	TOTAL
HEALTH	EXCELLENT	50	171	216	264	1	701
		12.3%	25.4%	33.1%	44.9%		30.2%
	GOOD	178	344	318	262	1	1102
		44.0%	51.1%	48.8%	44.6%		47.5%
	FAIR/POOR	177	158	118	62	4	515
		43.7%	23.5%	18.1%	10.5%		22.2%
	Missing	82	150	139	116	6	493
	TOTAL	405	673	652	588	12	2318
		100.0%	100.0%	100.0%	100.0%		

Education is very strongly related to people's health, as GSS respondents with a college degree are much more likely (44.9 percent) than those without a high school degree (12.3 percent) to report excellent health (V = .22**). Conversely, respondents without a high school degree are much more likely (43.7 percent) than those with a college degree (10.5 percent) to report only fair or poor health. Why do you think this relationship is so strong?

Recall from previous exercises that education is generally associated with lower acceptance of traditional beliefs and greater acceptance of nontraditional beliefs. We thus hypothesize that respondents with greater levels of education will be more likely than those with lower levels to oppose the death penalty.

Data File: **GSS**
Task: **Cross-tabulation**
➤ Row Variable: **47) EXECUTE?**
➤ Column Variable: **15) EDUCATION**
➤ View: **Tables**
➤ Display: **Column %**

EXECUTE?   by   EDUCATION
Cramer's V: 0.040

		EDUCATION					
		NOT H.S.	H.S. GRAD	SOME COLL.	COLL. GRAD	Missing	TOTAL
EXECUTE?	FAVOR	292	528	511	426	7	1757
		68.4%	69.8%	70.5%	65.7%		68.7%
	OPPOSE	135	228	214	222	2	799
		31.6%	30.2%	29.5%	34.3%		31.3%
	Missing	60	67	66	56	3	252
	TOTAL	427	756	725	648	12	2556
		100.0%	100.0%	100.0%	100.0%		

Our hypothesis is not supported: there is no relationship in this table between education and support for the death penalty (V = .04; n.s.). What might explain this surprising result? Recall from Exercise 5 that whites are much more likely than African Americans to support the death penalty. It is also true that whites overall have greater levels of education than African Americans. Perhaps a relationship between education and support for the death penalty in the table just presented is "masked" by the fact that African Americans are less likely to support the death penalty but also have lower levels of education. If so, we would expect to see the hypothesized result if we limit our analysis to whites.

Data File: **GSS**
Task: **Cross-tabulation**
Row Variable: **47) EXECUTE?**
Column Variable: **15) EDUCATION**
➤ Subset Variable: **20) RACE**
➤ Subset Category: **Include: 2) White**
➤ View: **Tables**
➤ Display: **Column %**

EXECUTE?   by   EDUCATION
Cramer's V: 0.095 **

		EDUCATION					
		NOT H.S.	H.S. GRAD	SOME COLL.	COLL. GRAD	Missing	TOTAL
EXECUTE?	FAVOR	230	462	447	378	5	1517
		76.2%	76.9%	77.1%	67.5%		74.3%
	OPPOSE	72	139	133	182	2	526
		23.8%	23.1%	22.9%	32.5%		25.7%
	Missing	43	48	51	46	0	188
	TOTAL	302	601	580	560	7	2043
		100.0%	100.0%	100.0%	100.0%		

Although the relationship is relatively weak, our hypothesis is now supported: whites with a college degree are about 9 percent less likely than those without a high school degree to support the death penalty (V = .10**).

RELIGION

Religion and Religious Belief in International Perspective

The NATIONS data set includes several variables dealing with religion. Let's map the distribution of two of the world's major religions, Christianity and Islam.

> *Data File:* **NATIONS**
> *Task:* **Mapping**
> *Variable 1:* **66) %CHRISTIAN**
> *View:* **Map**

%CHRISTIAN -- PERCENT CHRISTIAN (WCE)

This map shows the percent of a nation's population that is Protestant or Catholic. Christianity is obviously spread throughout the world, but is most common in the Western Hemisphere and parts of Europe.

Data File: **NATIONS**
Task: **Mapping**
> *Variable 1:* **65) %MUSLIM**
> *View:* **Map**

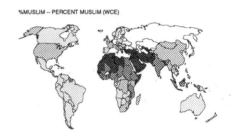

%MUSLIM -- PERCENT MUSLIM (WCE)

This map is pretty much the opposite of the one we just saw. The Muslim nations tend to be in northern Africa, the Middle East, and parts of Asia.

After the tragic events of September 11, 2001, Americans tried to understand why Muslim terrorists had targeted the United States. Without trying to excuse the horrible acts of that day, some observers noted that because most Muslim nations are much poorer than most Christian nations and thus have much worse living conditions, Muslims resent the United States and its Christian, Western European allies. We can illustrate the relationship between Christianity, Islam, and quality of life with two scatterplots.

Data File: **NATIONS**
> *Task:* **Scatterplot**
> *Dependent Variable:* **24) QUAL. LIFE**
> *Independent Variable:* **66) %CHRISTIAN**
> *View:* **Reg. Line**

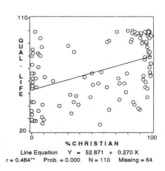

Line Equation   Y = 52.871 + 0.270 X
r = 0.464** Prob. = 0.000 N = 110 Missing = 64

Discovering Sociology

Here we see that the more Christian a nation, the higher its quality of life (r = .46**).

Data File: **NATIONS**
Task: **Scatterplot**
Dependent Variable: **24) QUAL. LIFE**
➤ Independent Variable: **65) %MUSLIM**
➤ View: **Reg. Line**

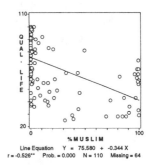

Line Equation   Y = 75.580 + -0.344 X
r = -0.526**   Prob. = 0.000   N = 110   Missing = 64

The opposite is true here, the more Muslim a nation, the lower its quality of life (r = −.53**). It is possible that the contrast between the poor living conditions in many Muslim nations and the better conditions in many Christian nations helped breed the resentment that exploded so horribly on September 11.

Religion in the United States

Your STATES data set contains some variables of interest for an understanding of religion and religiosity in the United States. We'll look at just a couple.

➤ Data File: **STATES**
➤ Task: **Mapping**
➤ Variable 1: **83) CH.MEMBERS**
➤ View: **Map**

CH.MEMBERS -- 1990: PERCENT OF POPULATION BELONGING TO A LOCAL CHURCH (CHURCH)

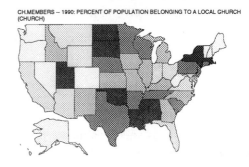

Church membership is lowest in the western part of the nation, with the notable exception of Utah. How would you explain this regional trend?

Another regional trend emerges when we examine the percent of each state's population that is Baptist.

Data File: **STATES**

Task: **Mapping**

➤ Variable 1: **33) % BAPTIST**

➤ View: **Map**

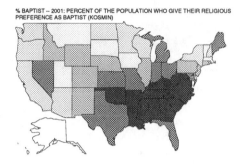

% BAPTIST -- 2001: PERCENT OF THE POPULATION WHO GIVE THEIR RELIGIOUS PREFERENCE AS BAPTIST (KOSMIN)

The South is clearly the most Baptist region in the country.

Let's turn now to the GSS. We'll first examine some ways in which the United States' three major religions differ from one another.

Let's see whether Protestants, Catholics, and Jews differ in their educational levels.

➤ Data File: **GSS**

➤ Task: **Cross-tabulation**

➤ Row Variable: **15) EDUCATION**

➤ Column Variable: **51) RELIGION**

➤ View: **Tables**

➤ Display: **Column %**

EDUCATION by RELIGION
Cramer's V: 0.110 **

	RELIGION				
	PROTESTANT	CATHOLIC	JEWISH	Missing	TOTAL
NOT H.S.	280	112	1	94	393
	18.5%	16.5%	1.6%		17.4%
H.S. GRAD	465	217	6	135	688
	30.7%	32.0%	9.7%		30.5%
SOME COLL.	426	197	18	150	641
	28.2%	29.0%	29.0%		28.4%
COLL. GRAD	342	153	37	172	532
	22.6%	22.5%	59.7%		23.6%
Missing	8	0	1	3	12
TOTAL	1513	679	62	554	2254
	100.0%	100.0%	100.0%		

Catholics and Protestants have similar levels of education while Jews have higher levels of education than the other two groups (V = .11**).

Do you think Catholics should be more likely than the other two groups to oppose abortion? As noted in Exercise 8, the GSS asks respondents whether they think a woman should be allowed to have a legal abortion if "she wants it for any reason."

Data File: **GSS**
Task: **Cross-tabulation**
➤ Row Variable: **90) ABORT ANY**
➤ Column Variable: **51) RELIGION**
➤ View: **Tables**
➤ Display: **Column %**

ABORT ANY by RELIGION
Cramer's V: 0.154 **

		RELIGION				
		PROTESTANT	CATHOLIC	JEWISH	Missing	TOTAL
ABORT ANY	YES	337	142	35	191	514
		35.2%	33.3%	76.1%		35.9%
	NO	621	284	11	147	916
		64.8%	66.7%	23.9%		64.1%
	Missing	563	253	17	216	1049
	TOTAL	958	426	46	554	1430
		100.0%	100.0%	100.0%		

Catholics do oppose abortion more than Jews, but so do Protestants (V = .15**); Catholics are not more likely than Protestants to oppose it.

Although we just saw that Catholics and Protestants do not differ in their education levels nor in their views about abortion, perhaps they will differ in their views about suicide. Since the Catholic Church officially condemns suicide as a sin, we hypothesize that Catholics should be less likely than Protestants to think that a person has a right to commit suicide if he or she has an incurable disease.

Data File: **GSS**
Task: **Cross-tabulation**
➤ Row Variable: **100) SUIC.ILL**
➤ Column Variable: **51) RELIGION**
➤ View: **Tables**
➤ Display: **Column %**

SUIC.ILL by RELIGION
Cramer's V: 0.122 **

		RELIGION				
		PROTESTANT	CATHOLIC	JEWISH	Missing	TOTAL
SUIC.ILL	YES	512	228	32	253	772
		52.8%	56.7%	91.4%		54.9%
	NO	457	174	3	111	634
		47.2%	43.3%	8.6%		45.1%
	Missing	552	277	28	190	1047
	TOTAL	969	402	35	554	1406
		100.0%	100.0%	100.0%		

Once again we see that Catholics and Protestants do not really differ; however, Jews are more likely than either of these two groups to support the right to suicide when someone has an incurable disease (V = .12**).

So far this exercise has considered education and religion as separate concepts. However, it is also true that education may be related to various religious views and behaviors. In particular, educational levels may be related to the likelihood of having certain religious beliefs and engaging in certain religious practices such as prayer. To illustrate this, we will examine whether education is related to religious beliefs about the Bible. The GSS asks respondents to choose one of the following: (a) the Bible is the actual word of God and is to be taken literally; (b) the Bible is the inspired word of God but not everything in it should be taken literally; and (c) the Bible is an ancient book of fables . . . recorded by men.

Data File: **GSS**

Task: **Cross-tabulation**

➤ Row Variable: **57) BIBLE**

➤ Column Variable: **15) EDUCATION**

➤ View: **Tables**

➤ Display: **Column %**

BIBLE   by   EDUCATION
Cramer's V: 0.184 **

		EDUCATION					
		NOT H.S.	H.S. GRAD	SOME COLL.	COLL. GRAD	Missing	TOTAL
BIBLE	ACTUAL	194	272	210	105	2	781
		51.9%	41.8%	32.8%	18.9%		35.1%
	INSPIRED	130	305	341	305	3	1081
		34.8%	46.9%	53.2%	54.9%		48.6%
	ANCI.BOOK	50	74	90	146	1	360
		13.4%	11.4%	14.0%	26.3%		16.2%
	Missing	113	172	150	148	6	589
	TOTAL	374	651	641	556	12	2222
		100.0%	100.0%	100.0%	100.0%		

Education is strongly related to a belief in the Bible as the actual word of God (V = .18**). People without a high school education are almost three times as likely as those with a college degree to believe the Bible is God's actual word.

WORKSHEET

NAME:

COURSE:

DATE:

EXERCISE

11

REVIEW QUESTIONS

Based on the first part of this exercise, answer True or False to the following items:

The higher a nation's educational level, the lower its opposition to abortion.	T	F
In the United States, college graduates tend to live in the Midwest.	T	F
In the GSS, the higher the education, the worse the health.	T	F
Africa is predominantly a Christian continent.	T	F
In the United States, church membership is low in the Midwest.	T	F
GSS data indicate that Catholics are more opposed to abortion than Protestants.	T	F

EXPLORIT QUESTIONS

1. In the preliminary section of this exercise, we saw lower educational levels linked to worse health. However, this relationship might be spurious once we take age into account, as older people may have lower levels of education and also worse health—not because of the lower educational levels but because of the effects of aging. Failing to control for age, then, may lead to an education-health relationship that exists only because of the effects of age on both education and health. To assess this possibility, we must control for age.

> ➤ *Data File:* **GSS**
> ➤ *Task:* **Cross-tabulation**
> ➤ *Row Variable:* **65) HEALTH**
> ➤ *Column Variable:* **15) EDUCATION**
> ➤ *Control Variable:* **14) AGE 65+**
> ➤ *View:* **Tables**
> ➤ *Display:* **Column %**

   a. Does a statistically significant relationship between education and health exist just among people ages 18–64?   Yes   No

   b. Does a statistically significant relationship between education and health exist just among people ages 65 and over?   Yes   No

   c. Is the original relationship between education and health spurious or not spurious?   Spurious   Not Spurious

2. Think of another variable in the GSS that might make the education-health relationship spurious once we control for it. It must be a variable that can affect both education and health. Use this as a control variable when you once again obtain a cross-tabulation between 65) HEALTH and 15) EDUCATION. Then answer the following questions.

    a. What variable (name and number) did you select as your control
        variable?   _____

    b. Why did you think this variable might make the original education-health relationship spurious? Explain your reasoning in detail.

    c. Examine your cross-tabulations and briefly summarize their results.

    d. Did the education-health relationship turn out to be spurious once you controlled for the variable you selected? Why or why not?

3. Education has long been regarded as a gateway to better-paying jobs. To illustrate this, we'll cross-tabulate education with a measure of occupation prestige where low prestige (under 40) is roughly equivalent to low-paying, blue-collar jobs and high prestige (40 and up) is roughly equivalent to higher-paying, white-collar jobs.

                    Data File: **GSS**
                        Task: **Cross-tabulation**
            ➤ Row Variable: **175) PRESTIGE**
         ➤ Column Variable: **15) EDUCATION**
                  ➤ View: **Tables**
               ➤ Display: **Column %**

    a. What percent of high school dropouts get high-prestige jobs?   _____%

    b. What percent of college graduates get high-prestige jobs?   _____%

    c. Do the results in this cross-tabulation prove that education is a pathway to better-paying jobs?

        1. Yes, because the relationship in the table between education and job prestige is so strong.

        2. No, because a third variable such as the education of the respondents' parents might make this relationship spurious.

4. Should states where many eighth-grade students watch 6 or more hours of TV per day have higher school dropout rates?

> ➤ *Data File:* **STATES**
> ➤ *Task:* **Scatterplot**
> ➤ *Dependent Variable:* **52) HS DROPOUT**
> ➤ *Independent Variable:* **84) %TV>6HRS**
> ➤ *View:* **Reg. Line**

    a. Use the results from the scatterplot to fill in the following sentence: The greater the percent of eighth-grade students watching 6 or more hours of TV per day in a state, the _____ the state's dropout rate.

    b. One possible conclusion to draw from this scatterplot is that if eighth-grade students watched TV less often, there would be _____ high-school dropouts.

    c. Locate California on the scatterplot and compare its position to the regression line. Does California have more high school dropouts than would normally be predicted from its amount of TV watching by eighth-graders, fewer dropouts, or about the expected number of dropouts? (Circle one.)

                                                      More

                                                     Fewer

                                             The expected number

5. Go into the NATIONS data set and obtain the rankings for the variable 116) EDUCATION. Then answer the following questions.

    a. Which country ranks highest (#1)?   _____

    b. What is this country's average years of schooling?   _____ years

    c. Which nation ranks lowest (#105)?   _____

    d. What is this nation's average years of schooling?   _____ years

6. The frequency of the use of contraception varies around the world. Let's see whether contraception use is higher in nations with more educated populations.

> Data File: **NATIONS**
> ➤ Task: **Scatterplot**
> ➤ Dependent Variable: **12) CONTRACEPT**
> ➤ Independent Variable: **116) EDUCATION**
> ➤ Display: **Reg. Line**

a. Which description best summarizes the results of this scatterplot?

     1. Nations with higher levels of education have greater levels of contraception use.

     2. Nations with higher levels of education have lower levels of contraception use.

     3. Nations with lower levels of education have greater levels of contraception use.

b. Does this scatterplot suggest that increasing the educational level of a nation might raise its use of contraception?   Yes   No

c. Use the [Find Case] option to locate the United States on the scatterplot. In which part of the scatterplot does the U.S. appear? (Circle one.)

         Upper left

         Upper right

         Lower left

         Lower right

7. The NATIONS data set includes a measure of the percent of each nation's sample who describe themselves as "a religious person."

> Data File: **NATIONS**
> Task: **Scatterplot**
> ➤ Dependent Variable: **7) BIRTH RATE**
> ➤ Independent Variable: **71) REL.PERSON**
> ➤ View: **Reg. Line**

a. Which description best summarizes the results of this scatterplot?

     1. Nations with lower levels of religiosity have higher birth rates.

     2. Nations with higher levels of religiosity have lower birth rates.

     3. Nations with higher levels of religiosity have higher birth rates.

b. Locate Nigeria on the scatterplot and compare its position to the regression line. Does Nigeria have a higher birth rate than would normally be predicted from its level of religiosity, a lower birth rate, or about the expected birth rate? (Circle one.)

         Higher

         Lower

         The expected birth rate

8. People who believe the Bible is the actual word of God and take the Bible literally are often referred to as religious fundamentalists. Let's explore whether this type of religious belief is linked to other beliefs.

> ➤ *Data File:* **GSS**
> ➤ *Task:* **Cross-tabulation**
> ➤ *Row Variable:* **98) SPANKING**
> ➤ *Column Variable:* **57) BIBLE**
> ➤ *View:* **Tables**
> ➤ *Display:* **Column %**

a. What percent of people who take the Bible literally approve of spanking?   _____%

b. What percent of people who think the Bible is an ancient book of fables
   approve of spanking?   _____%

c. Is V statistically significant?   Yes   No

d. Is belief in the Bible linked to views about spanking?   Yes   No

9. Now let's consider abortion attitudes.

> *Data File:* **GSS**
> *Task:* **Cross-tabulation**
> ➤ *Row Variable:* **90) ABORT ANY**
> ➤ *Column Variable:* **57) BIBLE**
> ➤ *View:* **Tables**
> ➤ *Display:* **Column %**

a. Fill in the following sentence: People who take the Bible literally are _____ percent more likely to oppose abortion than those who think the Bible is an ancient book of fables.

b. More than half of the people who believe the Bible is an ancient book of fables
   also believe that abortion should be made legal.   T   F

10. What about beliefs regarding women's role in society?

> *Data File:* **GSS**
> *Task:* **Cross-tabulation**
> ➤ *Row Variable:* **105) WIFE@HOME**
> ➤ *Column Variable:* **57) BIBLE**
> ➤ *View:* **Tables**
> ➤ *Display:* **Column %**

Taking into account V, do the results of this cross-tabulation suggest that religious fundamentalism is related to beliefs about women's role in society?   Yes   No

11. Should fundamentalists be especially likely to think that preschool children suffer if their mothers work?

> *Data File:* **GSS**
> *Task:* **Cross-tabulation**
> ➤ *Row Variable:* **104) PRESCH.WRK**
> ➤ *Column Variable:* **57) BIBLE**
> ➤ *View:* **Tables**
> ➤ *Display:* **Column %**

a. What percent of people who take the Bible literally think such children suffer?   _____%

b. What percent of people who think the Bible is an ancient book of fables think these children suffer?   _____%

c. Is V statistically significant?   Yes   No

d. Is belief in the Bible linked to beliefs about whether preschool children suffer if their mothers work?   Yes   No

e. More than half of the people who think the Bible is the actual word of God also believe that preschool children suffer if their mothers work.   T   F

12. In the United States, Protestantism consists of several denominations that differ not only in their religious beliefs but also in social characteristics. Let's see whether they differ in their views about abortion.

> *Data File:* **GSS**
> *Task:* **Cross-tabulation**
> ➤ *Row Variable:* **90) ABORT ANY**
> ➤ *Column Variable:* **164) DENOM**
> ➤ *View:* **Tables**
> ➤ *Display:* **Column %**

a. Which denomination is most in favor of legalized abortion for any reason (i.e., a "yes" response)?   _____

b. Which denomination is least likely to give a "yes" response?   _____

c. More than half of Lutherans believe that abortion should not be made legal.   T   F

◆ EXERCISE 12 ◆

WORK AND THE ECONOMY

Tasks: Mapping, Scatterplot, Univariate, Cross-tabulation
Data Files: NATIONS, STATES, GSS

Every society has an *economy*, or a system for the production and distribution of goods and services. The type of economy a society enjoys has important implications for the lives of its members. In today's world, agricultural societies are much poorer than industrial ones and, as a result, have lower life expectancies, higher rates of disease, and other such problems. A society's economy also affects its members' views on many issues, including those not related to work or the economy.

In the United States, sociologists, economists, and other scholars have carried out much research on work and employment. Many studies have examined the sources and consequences of unemployment, the determinants of job satisfaction, the aspects of work that Americans think are most important, and other matters. This body of research has yielded a rich understanding of work in U.S. society.

This exercise, then, examines work and economy in the United States and around the world. We begin by looking internationally at agricultural economies and at their implications for the lives of individuals. We then turn to the United States to explore such issues as unemployment and job satisfaction.

INTERNATIONAL ECONOMIES AND LIFE CHANCES

As noted above, many of today's nations continue to be primarily agricultural. Let's map the percent of gross domestic product (GDP) that is accounted for by agriculture to see where the agricultural nations lie.

> ➤ *Data File:* **NATIONS**
> ➤ *Task:* **Mapping**
> ➤ *Variable 1:* **34) % AGRIC $**
> ➤ *View:* **Map**

% AGRIC $ -- PERCENT OF GDP [GROSS DOMESTIC PRODUCT] ACCOUNTED FOR BY AGRICULTURE (TWF, 1997)

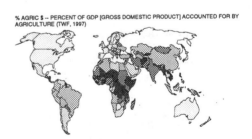

The most agriculturally dependent nations lie in Africa and parts of Asia, the least in North America and Europe.

Data File:	**NATIONS**		
Task:	**Mapping**		
Variable 1:	**34) % AGRIC $**		
➤ *View:*	**List: Rank**		

RANK	CASE NAME	VALUE
1	Georgia	70.4
2	Somalia	65.5
3	Myanmar	63.0
4	Congo, Dem. Republic	59.0
5	Ethiopia	57.0
5	Tanzania	57.0
7	Albania	56.0
7	Burundi	56.0
7	Laos	56.0
7	Afghanistan	56.0

In terms of GDP, the most agricultural nation is Georgia, where 70.4 percent of the GDP is accounted for by agriculture. The least agricultural nations are Singapore and Kuwait, where none of the GDP is agriculturally based. The United States ranks just above that, at 2 percent.

Another variable for mapping the agricultural nations is the percent of the labor force employed by agriculture.

Data File:	**NATIONS**
Task:	**Mapping**
➤ *Variable 1:*	**37) %WORK AG**
➤ *View:*	**Map**

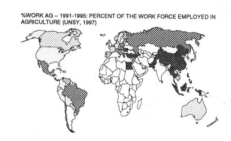

%WORK AG -- 1991-1995: PERCENT OF THE WORK FORCE EMPLOYED IN AGRICULTURE (UNSY, 1997)

Not surprisingly, this map looks very similar to the previous one.

If the poorest nations in the world are agricultural, the agricultural nations should be the worst off in terms of life expectancy, illness, and other problems. Let's find out.

Data File:	**NATIONS**
➤ *Task:*	**Scatterplot**
➤ *Dependent Variable:*	**18) LIFE EXPCT**
➤ *Independent Variable:*	**37) %WORK AG**
➤ *View:*	**Reg. Line**

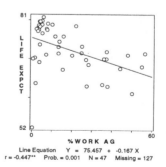

Line Equation   Y = 75.457 + -0.167 X
r = -0.447**   Prob. = 0.001   N = 47   Missing = 127

This is a strong correlation (r = −.45**). The most agricultural nations have lower life expectancies than the least agricultural ones. The latter, of course, are the world's industrial societies.

What about infant mortality?

Data File: **NATIONS**
Task: **Scatterplot**
➤ Dependent Variable: **10) INF.MORTL**
➤ Independent Variable: **37) %WORK AG**
➤ View: **Reg. Line**

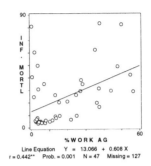

The agricultural nations have higher levels of infant mortality (r = .44**).

Data File: **NATIONS**
Task: **Scatterplot**
➤ Dependent Variable: **116) EDUCATION**
➤ Independent Variable: **37) %WORK AG**
➤ View: **Reg. Line**

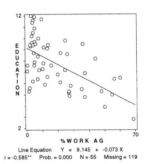

And the agricultural nations have much lower levels of education (r = −.59**).

WORK AND EMPLOYMENT IN THE UNITED STATES

Do you know which states are the most agricultural? Let's examine this and other issues using the STATES data file.

> *Data File:* **STATES**
> *Task:* **Mapping**
> *Variable 1:* **85) %AGRI.EMP**
> *View:* **Map**

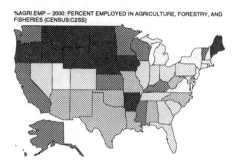

%AGRI.EMP -- 2000: PERCENT EMPLOYED IN AGRICULTURE, FORESTRY, AND FISHERIES (CENSUS:C2SS)

The agricultural, forestry, and fishery states tend to be in the upper Midwest.

At the international level, the agricultural nations are poor. In the United States, are the more agricultural states also poorer?

> *Data File:* **STATES**
> *Task:* **Scatterplot**
> *Dependent Variable:* **47) PER CAP$**
> *Independent Variable:* **85) %AGRI.EMP**
> *View:* **Reg. Line**

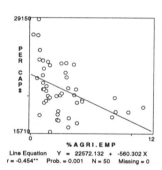

Line Equation   Y = 22572.132 + -560.302 X
r = -0.454**   Prob. = 0.001   N = 50   Missing = 0

The relationship is statistically significant (r = −.45**); per capita income is lower in the more agricultural states.

One of the toughest things about work in the United States is getting to your job in the morning and getting home at night. Many people take public transportation—a bus, train, cab, subway, and so on. Let's see which states have the highest use of public transportation.

> *Data File:* **STATES**
> *Task:* **Mapping**
> *Variable 1:* **86) PUB. TRANS.**
> *View:* **Map**

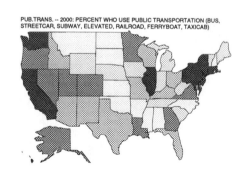

PUB.TRANS. -- 2000: PERCENT WHO USE PUBLIC TRANSPORTATION (BUS, STREETCAR, SUBWAY, ELEVATED, RAILROAD, FERRYBOAT, TAXICAB)

Public transportation is most often used in New York, New Jersey, Massachusetts, and several eastern states.

What accounts for this range in public transportation use? The obvious answer is urbanism: in rural areas, people are more likely to drive to work in their cars or pickups, and public transportation often doesn't exist in the first place. Let's see whether the more urban states have higher public transportation use.

Data File: **STATES**
➤ Task: **Scatterplot**
➤ Dependent Variable: **86) PUB. TRANS**
➤ Independent Variable: **19) %URBAN**
➤ View: **Reg. Line**

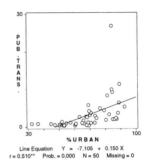

Bingo! The correlation, r, is .51**.

If you take public transportation to work, does that take more time or less time than if you drive to work or get there some other way (e.g., walking or bicycling)? Our dependent variable will be the average time it takes to get to work.

Data File: **STATES**
Task: **Scatterplot**
➤ Dependent Variable: **87) AVG. TRAVL**
➤ Independent Variable: **86) PUB. TRANS**
➤ View: **Reg. Line**

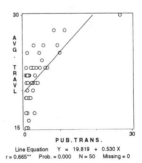

The states with the highest public transportation use tend to be the states where it takes the most time to get to work ($r = .66$**). Note, however, that this does not *prove* public transportation takes more time than driving. If you lived in New York City or another big city, public transportation could take less time than driving would. All the scatterplot suggests is that those who do take public transportation spend more time commuting than those who drive or use some other means to get to work. What makes the most sense for any one individual obviously depends heavily on where that person lives. There is also the question of which type of commuting would cause the least pollution—but that's another question altogether.

Time now for the GSS. One of the most important aspects of a job is how much you like it. The GSS asks whether respondents are satisfied with the work they do.

➤ *Data File:* **GSS**
➤ *Task:* **Univariate**
➤ *Primary Variable:* **174) LIKE JOB?**
➤ *View:* **Pie**

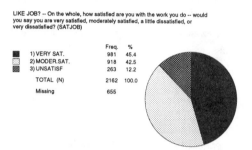

LIKE JOB? -- On the whole, how satisfied are you with the work you do -- would you say you are very satisfied, moderately satisfied, a little dissatisfied, or very dissatisfied? (SATJOB)

	Freq.	%
1) VERY SAT.	981	45.4
2) MODER.SAT.	918	42.5
3) UNSATISF	263	12.2
TOTAL (N)	2162	100.0
Missing	655	

Forty-five percent of the respondents are very satisfied with their jobs, 43 percent are moderately satisfied, and 12 percent are dissatisfied. What affects job satisfaction? At least since the time of Karl Marx, social scientists have observed that blue-collar jobs are less creative and more alienating than white-collar jobs. If so, people working in the latter should be more satisfied with their jobs than people working in the former. Let's see if this is so, using the GSS measure of occupational prestige that roughly corresponds to the white collar–blue collar distinction.

Data File: **GSS**
➤ *Task:* **Cross-tabulation**
➤ *Row Variable:* **174) LIKE JOB?**
➤ *Column Variable:* **175) PRESTIGE**
➤ *View:* **Tables**
➤ *Display:* **Column %**

LIKE JOB? by PRESTIGE
Cramer's V: 0.087 **

		PRESTIGE			
		< 40	40 AND UP	Missing	TOTAL
LIKE JOB?	VERY SAT.	328	619	34	947
		40.0%	48.8%		45.3%
	MODER.SAT.	380	505	33	885
		46.3%	39.8%		42.3%
	UNSATISF	113	145	5	258
		13.8%	11.4%		12.3%
	Missing	261	309	85	655
	TOTAL	821	1269	157	2090
		100.0%	100.0%		

People in high-prestige jobs are slightly more likely (48.8 percent) than those in low-prestige ones (40 percent) to be very satisfied with their jobs (V = .09**).

Although this result makes sense, we still have to consider the possibility that it might be spurious. For example, African Americans generally have lower-prestige jobs than whites and may also, for different reasons such as racial discrimination in the workplace, may like their jobs less than whites do. We therefore will control for 20) RACE to test for spuriousness.

Data File: **GSS**

Task: **Cross-tabulation**

Row Variable: **174) LIKE JOB?**

Column Variable: **175) PRESTIGE**

➤ Control Variable: **20) RACE**

➤ View: **Tables**

➤ Display: **Column %**

		PRESTIGE			
		< 40	40 AND UP	Missing	TOTAL
LIKE JOB?	VERY SAT.	48	74	5	122
		33.3%	43.5%		38.9%
	MODER.SAT.	70	67	10	137
		48.6%	39.4%		43.6%
	UNSATISF	26	29	2	55
		18.1%	17.1%		17.5%
	Missing	63	21	17	101
	TOTAL	144	170	34	314
		100.0%	100.0%		

When we do so, we find that the original relationship between occupational prestige and job satisfaction continues to be statistically significant for whites (V = .07*) but not for African Americans (V = .11; n.s.), even though African Americans in higher-prestige jobs are about 8 percent more likely than those in low-prestige jobs to be very satisfied with their work. Although this percentage difference is sizable, the number of African Americans in the entire table is too small to allow it to be statistically significant. Still, the percentage differences we do find, coupled with the statistically significant relationship between occupational prestige and job satisfaction among whites, give us some confidence that the original relationship between prestige and job satisfaction was *not* spurious.

Compared to men, women are more likely to be in low-paying jobs in the service sector and are more likely to be sexually harassed in the workplace. Does that mean they should be less satisfied than men with their jobs? We'll limit our analysis to full-time workers.

Data File: **GSS**

Task: **Cross-tabulation**

Row Variable: **174) LIKE JOB?**

➤ Column Variable: **19) GENDER**

➤ Subset Variable: **1) FULL TIME?**

➤ Subset Category: **Include: 1) Yes**

➤ View: **Tables**

➤ Display: **Column %**

LIKE JOB? by GENDER

Cramer's V: 0.079 **

		GENDER		
		FEMALE	MALE	TOTAL
LIKE JOB?	VERY SAT.	311	368	679
		43.5%	45.9%	44.8%
	MODER.SAT.	306	364	670
		42.8%	45.4%	44.2%
	UNSATISF	98	70	168
		13.7%	8.7%	11.1%
	Missing	3	4	7
	TOTAL	715	802	1517
		100.0%	100.0%	

Women are slightly more likely than men to be unsatisfied with their jobs (V = .08**). This finding does not imply that women avoid the workplace problems just mentioned. Evidently, when they answer the GSS question, they think of other aspects of their jobs.

If educated people have higher-prestige jobs, as we saw in Exercise 11, should college-educated people express higher job satisfaction than people without a high school degree?

Data File: **GSS**
Task: **Cross-tabulation**
Row Variable: **174) LIKE JOB?**
➤ Column Variable: **15) EDUCATION**
➤ View: **Tables**
➤ Display: **Column %**

LIKE JOB? by EDUCATION
Cramer's V: 0.062 *

		EDUCATION					
		NOT H.S.	H.S. GRAD	SOME COLL.	COLL. GRAD	Missing	TOTAL
LIKE JOB?	VERY SAT.	122	246	313	298	2	979
		42.2%	39.6%	48.7%	49.5%		45.4%
	MODER.SAT.	129	291	255	238	5	913
		44.6%	46.9%	39.7%	39.5%		42.4%
	UNSATISF	38	84	75	66	0	263
		13.1%	13.5%	11.7%	11.0%		12.2%
	Missing	198	202	148	102	5	655
	TOTAL	289	621	643	602	12	2155
		100.0%	100.0%	100.0%	100.0%		

The relationship is statistically significant but rather weak (V = .06*). Did you expect a stronger relationship?

Let's change our focus to unemployment. A vast amount of literature documents the psychological and other effects of unemployment on individuals and their families. We can explore a few of these effects with GSS data by comparing people who are currently unemployed with those who are not unemployed. For now, let's look just at happiness.

Data File: **GSS**
Task: **Cross-tabulation**
➤ Row Variable: **63) HAPPY?**
➤ Column Variable: **4) UNEMPLOYED**
➤ View: **Tables**
➤ Display: **Column %**

HAPPY? by UNEMPLOYED
Cramer's V: 0.088 **

		UNEMPLOYED		
		YES	NO	TOTAL
HAPPY?	VERY HAPPY	16	865	881
		27.1%	31.8%	31.7%
	PRET.HAPPY	26	1577	1603
		44.1%	58.0%	57.7%
	NOT TOO	17	276	293
		28.8%	10.2%	10.6%
	Missing	1	39	40
	TOTAL	59	2718	2777
		100.0%	100.0%	

Unemployed respondents are almost three times as likely as those who are not unemployed to say they're not too happy (V = .09**).

We hear a lot these days about affirmative action, which is usually interpreted as preferential hiring given to racial minorities to redress past discrimination and injustice. Opponents of affirmative action term it "reverse discrimination" against qualified whites. The GSS asked respondents how likely they think it is "that a white person won't get a job or promotion while an equally or less qualified black person gets one instead." How would you respond to this question?

Discovering Sociology

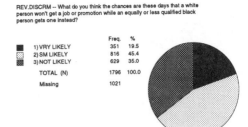

REV.DISCRM -- What do you think the chances are these days that a white person won't get a job or promotion while an equally or less qualified black person gets one instead?

	Freq.	%
■ 1) VRY LIKELY	351	19.5
▨ 2) SM LIKELY	816	45.4
▩ 3) NOT LIKELY	629	35.0
TOTAL (N)	1796	100.0
Missing	1021	

Data File: **GSS**
➤ Task: **Univariate**
➤ Primary Variable: **124) REV.DISCRM**
➤ View: **Pie**

Opinion is pretty well divided. Almost 20 percent say the chances of a white person not getting a job under these circumstances are very likely, about 45 percent say somewhat likely, and 35 percent say not likely.

What affects our views about affirmative action (or reverse discrimination)? It makes sense to think that whites and African Americans should have very different views. Our hypothesis is that whites will be more apt than African Americans to think that the chances of a white person not getting a job are very likely and less apt to rate the chances as not likely.

Data File: **GSS**
➤ Task: **Cross-tabulation**
➤ Row Variable: **124) REV.DISCRM**
➤ Column Variable: **20) RACE**
➤ View: **Tables**
➤ Display: **Column %**

REV.DISCRM   by   RACE
Cramer's V: 0.191 **

		RACE			
		BLACK	WHITE	Missing	TOTAL
R E V . D I S C R M	VRY LIKELY	37	301	13	338
		14.3%	20.9%		19.9%
	SM LIKELY	77	696	43	773
		29.7%	48.3%		45.5%
	NOT LIKELY	145	443	41	588
		56.0%	30.8%		34.6%
	Missing	173	798	50	1021
	TOTAL	259	1440	147	1699
		100.0%	100.0%		

The two races certainly have different views, and our hypothesis is clearly supported (V = .19**).

NAME:

COURSE:

DATE:

REVIEW QUESTIONS

Based on the first part of this exercise, answer True or False to the following items:

In no country of the world is more than 60 percent of the GDP accounted for by agriculture. T F

Agricultural nations have higher infant mortality rates than industrial nations. T F

In the United States, the southeastern states have the highest proportion of people employed in agriculture. T F

In the United States, a state's degree of "agriculturalism" is not related to its level of income. T F

In the GSS, women are less satisfied than men with their jobs. T F

In the GSS, unemployed people feel less happy than working people. T F

EXPLORIT QUESTIONS

1. In the last example of the preliminary section, you discovered a fairly strong racial difference in views on the likelihood of a white person not getting a job while a white person does. Should we also expect to find a racial difference in the percent who favor "preference in hiring and promotion"?

> ➤ *Data File:* **GSS**
> ➤ *Task:* **Cross-tabulation**
> ➤ *Row Variable:* **61) AFFRM. ACT**
> ➤ *Column Variable:* **20) RACE**
> ➤ *View:* **Tables**
> ➤ *Display:* **Column %**

a. Complete the following sentence: Black respondents are _____ percent more likely than white respondents to favor affirmative action.

b. Do the results of this cross-tabulation support the view (from conflict theory) that people favor policies according to their own vested interests? Yes No

2.   Preference in hiring can be based on gender as well as on race. Since we have found a racial differ-
     ence in support for affirmative action, do you think we will also find a gender difference?

                          Data File: **GSS**
                              Task: **Cross-tabulation**
                     Row Variable: **61) AFFRM. ACT**
        ➤ *Column Variable:* **19) GENDER**
                    ➤ *View:* **Tables**
                 ➤ *Display:* **Column %**

     a. Is V statistically significant?   Yes   No

     b. Taking into account V, is gender related to approval for affirmative action in hiring?   Yes   No

     c. More than one-fourth of female respondents favor affirmative action.   T   F

3.   Select another independent variable that might predict support for affirmative action as measured by
     61) AFFRM. ACT. Obtain the appropriate cross-tabulation and then answer these questions.

     a. What variable (name and number) did you select?   _____

     b. What is your hypothesis?

     c. Why do you think this hypothesis will be supported?

     d. Briefly summarize the results of your cross-tabulation and whether they support your hypothesis.

4. In the preliminary section of this exercise, you used the GSS to examine the relationship between happiness and one's employment status. At the international level, let's see if happiness is lower in nations with higher unemployment.

> ➤ *Data File:* **NATIONS**
> ➤ *Task:* **Scatterplot**
> ➤ *Dependent Variable:* **110) VERY HAPPY**
> ➤ *Independent Variable:* **31) UNEMPLYRT**
> ➤ *View:* **Reg. Line**

a. What is the value of r?   r = _____

b. Is r statistically significant?   Yes   No

c. Which statement below best describes the results of this scatterplot?
    1. The higher the unemployment rate, the happier a nation's people.
    2. The lower the unemployment rate, the happier a nation's people.
    3. There is no relationship between the unemployment rate and happiness.

5. The GSS asks respondents whether they agree that "family life often suffers because men concentrate too much on their work." The variable is 136) MEN.OVRWRK. Because this variable implicitly blames men for why family life can suffer, we will hypothesize that men will be less likely than women to agree with this statement. To test this hypothesis, obtain a cross-tabulation where 136) MEN. OVRWRK is the dependent (row) variable and 19) GENDER is the independent (column) variable.

a. What percent of women agree with this statement?   _____ %

b. What percent of men agree with this statement?   _____ %

c. Taking into account V and the percent you just listed, is the hypothesis
    supported?   Yes   No

6. The NATIONS data set includes a variable on the percent of children who are seriously underweight. We have seen that people in agricultural nations have some very negative life chances. Are their children especially likely to be underweight? To find out, obtain a scatterplot where 27) %UNDRWGHT is the dependent variable and 37) %WORK AG is the independent variable.

a. Which description below best describes the results of this scatterplot?
    1. The less agricultural a nation, the higher its proportion of underweight children.
    2. The more agricultural a nation, the higher its proportion of underweight children.
    3. The more agricultural a nation, the lower its proportion of underweight children.

b. Click on the dot at the top right of the scatterplot to determine which
    nation both is the most agricultural and has the highest proportion of
    underweight children. What is the name of this nation?   _____

Exercise 12: Work and the Economy

c. Using all the results you've obtained, write a brief essay below in which you describe the distribution of underweight children throughout the world and comment on the question of whether they are especially likely to be found in agricultural nations.

7. Have you ever wondered what life would be like if you won a state lottery? In particular, have you wondered whether you would continue to work if you didn't have to? The GSS asks respondents, "If you were to get enough money to live as comfortably as you would like for the rest of your life, would you continue to work or would you stop working?" This is a rough measure of the commitment to the so-called "work ethic." Obtain a cross-tabulation where 83) WORK IF $$ is the dependent (row) variable and 20) RACE is the independent (column) variable. This cross-tabulation will allow us to see whether there are racial differences in commitment to the work ethic.

a. What percent of black respondents say they would continue to work? _____%

b. What percent of white respondents say they would continue to work? _____%

c. Is V statistically significant?   Yes   No

d. Are there racial differences in commitment to the work ethic as measured by the dependent variable?   Yes   No

8. Do you think gender will affect whether people would want to work?

a. Let's see whether your hypothesis was correct. First obtain the following cross-tabulation.

> Data File: **GSS**
> ➤ Task: **Cross-tabulation**
> ➤ Row Variable: **83) WRK IF $$**
> ➤ Column Variable: **19) GENDER**
> ➤ View: **Tables**
> ➤ Display: **Column %**

Now answer these questions.

b. What percent of women say they would continue to work? _____%

c. What percent of men? _____%

d. Is V statistically significant?   Yes   No

e. Did the data support your hypothesis?   Yes   No

9. Next, let's see whether family income predicts whether people would want to work if they didn't have to for economic reasons.

> Data File: **GSS**
> Task: **Cross-tabulation**
> Row Variable: **83) WRK IF $$**
> ➤ Column Variable: **26) FAM INCOME**
> ➤ View: **Tables**
> ➤ Display: **Column %**

a. What percent of people with annual incomes under $25,000 would continue to work?   _____%

b. What percent of people with annual incomes $50,000 and over would continue to work?   _____%

c. Is V statistically significant?   Yes   No

d. Do these results suggest that the stereotype that poor people are lazy is incorrect?   Yes   No

10. The NATIONS data set includes a variable measuring the percent of a nation's population who say that work is "very important" in their lives.

> ➤ Data File: **NATIONS**
> ➤ Task: **Mapping**
> ➤ Variable 1: **85) WORK IMPT?**
> ➤ Display: **Map**

a. Which of the following regions seems most likely to believe that work is important in the lives of its residents? (Circle one.)

North America

South America

Europe

b. What percent of people in the United States say that work is very important in their lives?   _____%

c. What is the ranking of the United States on this variable?   _____th

11. We will next examine the relationship between a nation's educational level and the percent of its population who feel that work is very important in their lives.

    a. What hypothesis do you think this relationship will support? Why?

> Data File: **NATIONS**
> ➤ Task: **Scatterplot**
> ➤ Dependent Variable: **85) WORK IMPT?**
> ➤ Independent Variable: **116) EDUCATION**
> ➤ Display: **Reg. Line**

    b. Complete this sentence: The higher the educational level of a nation, the _____ the percent of its people who say that work is very important in their lives.

    c. Locate the United States in the scatterplot. In which part of the scatterplot is it found? (Circle one.)

Upper left

Upper right

Lower left

Lower right

    d. This scatterplot suggests that the United State ranks relatively low on the 85) WORK IMPT? variable because it has such a high level of education.   T   F

POLITICS AND GOVERNMENT

Tasks: Mapping, Cross-tabulation, Scatterplot, Historical Trends, Univariate
Data Files: NATIONS, STATES, GSS, HISTORY

Nations around the world differ in the types of government that run them. Some, like the United States, are democracies in which people elect their representatives and political freedom is generally the rule. Others are more authoritarian; in these, elections are rare or are "sham" elections if they do occur, and one person or a small group of self-appointed individuals rules the government. Political freedom is uncommon and, even worse, political repression the norm; people live in fear for their lives if they dare question the arbitrary power under which they live.

Some of the most important questions in the study of politics and society concern the reasons why ordinary citizens become involved, or fail to become involved, in politics. In democracies, the most common political activity is voting, on which many studies exist. As we'll see, some types of people are more likely to vote than others. In both democratic and authoritarian nations, citizens often try to influence the political process through nonelectoral means, including protest. Many types of protest exist, and we'll be looking at a few of them.

Our exploration begins with a look at political structures, activities, and attitudes across the world and within the United States. As you read through this exercise, think about why you have the attitudes you do about politics and government and why you've voted or engaged in some other political activity or, perhaps, failed to vote or otherwise become involved in politics. The exercise will give you a better idea of the social forces affecting your political behavior and attitudes.

GLOBAL POLITICS

The world's nations have been ranked according to how much political freedom their citizens enjoy. In the freest or most democratic nations, people vote via a secret ballot and enjoy freedom of speech, other political freedoms, and civil liberties. In the least free or most authoritarian nations, none of these hallmarks of democracy exist. In the NATIONS data set, the most democratic nations get a score of 7 and the most authoritarian get a score of 1. Let's see which nations are the most democratic and which are the most authoritarian.

> *Data File:* **NATIONS**
> *Task:* **Mapping**
> *Variable 1:* **54) FREEDOM**
> *View:* **Map**

FREEDOM – FREEDOM IN THE WORLD OVERALL RATING AS AVERAGE OF
POLITICAL RIGHTS AND CIVIL LIBERTIES: 1 = LEAST FREE, 7 = MOST FREE (FITW

The darker the color, the more democratic. As you might have expected, the most democratic nations tend to be in North America and Western Europe.

After the terrorist attacks on September 11, 2001, many news articles appeared about life and politics in Muslim societies. They typically pointed out the authoritarian nature of government in many of these nations. We can illustrate this trend with a scatterplot.

Data File: **NATIONS**

➤ Task: **Scatterplot**

➤ Dependent Variable: **54) FREEDOM**

➤ Independent Variable: **65) %MUSLIM**

➤ View: **Reg. Line**

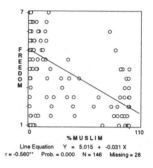

Line Equation Y = 5.015 + -0.031 X
r = -0.560** Prob. = 0.000 N = 146 Missing = 28

Generally speaking, the more Muslim a nation's population, the less freedom its citizens enjoy (r = −.56**).

An important variable to political sociologists is political interest, which, as the name implies, concerns the amount of interest someone has in politics. The NATIONS data set includes a measure of the percent of each nation's respondents who are very interested or somewhat interested in politics.

Data File: **NATIONS**

➤ Task: **Mapping**

➤ Variable 1: **60) P.INTEREST**

➤ View: **Map**

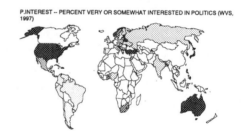

P.INTEREST -- PERCENT VERY OR SOMEWHAT INTERESTED IN POLITICS (WVS, 1997)

The amount of political interest varies around the world, and no regional patterns are obvious.

We said earlier that people sometimes take part in nonelectoral activity to influence the political process. The NATIONS data set includes three such activities: signing a petition, taking part in a boycott, and participating in a lawful demonstration.

		Data File:	**NATIONS**
		Task:	**Mapping**
➤		Variable 1:	**61) PETITION?**
	➤	View:	**List: Rank**

RANK	CASE NAME	VALUE
1	Australia	78.9
2	Sweden	71.6
3	United States	71.2
4	Switzerland	68.0
5	Germany	65.9
6	Norway	64.7
7	United Kingdom	57.6
8	Japan	55.0
9	Brazil	47.1
10	Croatia	42.7

The percent of citizens who say they've signed a political petition ranges from a high of 78.9 percent in Australia to a low of only 6.7 percent in Nigeria.

POLITICS IN THE UNITED STATES

The 2000 presidential election seemed to last forever, as the results were too close to call for several weeks thanks to a virtual deadlock in Florida. Republican George W. Bush was eventually declared the winner over democrat Al Gore. Let's see where in the country Bush's support was highest and see whether we can determine, demographically speaking, what types of states gave Bush the most and the least support.

> ➤ *Data File:* **STATES**
> ➤ *Task:* **Mapping**
> ➤ *Variable 1:* **69) STATES 00**
> ➤ *View:* **Map**

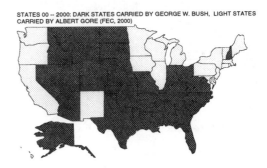

STATES 00 -- 2000: DARK STATES CARRIED BY GEORGE W. BUSH, LIGHT STATES CARRIED BY ALBERT GORE (FEC, 2000)

Bush carried the dark states and Gore won the light states. Bush generally won the South, much of the Midwest, and mountain states, while Gore won the Northeast, the West Coast, and some of the Midwest.

	Data File:	**STATES**
	Task:	**Mapping**
➤	Variable 1:	**70) %GWBUSH00**
	➤ View:	**List: Rank**

RANK	CASE NAME	VALUE
1	Wyoming	67.8
2	Idaho	67.2
3	Utah	66.8
4	Nebraska	62.3
5	North Dakota	60.7
6	Oklahoma	60.3
6	South Dakota	60.3
8	Texas	59.3
9	Alaska	58.6
10	Montana	58.4

The percentage actually voting for Bush was highest in the mountain states, Texas, and Oklahoma.

Why were some states likely to favor Bush while others were likely to favor Gore? Historically, the Democratic Party has been seen as the party of the working class, whereas the Republican Party has been seen as the party of the wealthy. If that's the case, states with higher median family income should have been more likely to vote for Bush.

	Data File:	**STATES**
	➤ Task:	**Scatterplot**
➤	Dependent Variable:	**70) %GWBUSH00**
➤	Independent Variable:	**46) MED.FAM. $**
	➤ View:	**Reg. Line**

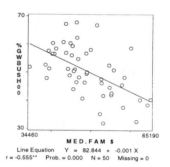

Line Equation   Y = 82.844 + -0.001 X
r = -0.555**   Prob. = 0.000   N = 50   Missing = 0

The scatterplot is the opposite of what we expected (r = −.55**)! How would you explain this surprising result?

Since the presidency of Franklin Delano Roosevelt some 60 years ago, African Americans have supported the Democratic Party in great numbers. Perhaps the states with the highest proportions of African Americans were less likely to vote for Bush.

Data File: **STATES**
Task: **Scatterplot**
Dependent Variable: **70) %GWBUSH00**
➤ Independent Variable: **14) %BLACK**
➤ View: **Reg. Line**

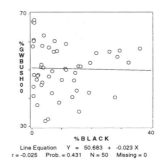

Line Equation   Y = 50.683 + -0.023 X
r = -0.025   Prob. = 0.431   N = 50   Missing = 0

No relationship exists between the two variables (r = n.s.).

We now turn to the GSS and continue our look at voting. Let's first get an idea of what kinds of people were more and less likely to vote in 1996. We'll start with education. One of the most consistent findings in the voting literature is that more-educated people vote more regularly than less-educated people. Let's find out whether the GSS data reflect this difference.

➤ Data File: **GSS**
➤ Task: **Cross-tabulation**
➤ Row Variable: **29) VOTE IN 96**
➤ Column Variable: **15) EDUCATION**
➤ View: **Tables**
➤ Display: **Column %**

VOTE IN 96   by   EDUCATION
Cramer's V: 0.236 **

		EDUCATION					
		NOT H.S	H.S. GRAD	SOME COLL	COLL. GRAD	Missing	TOTAL
VOTE IN 96	VOTED	195	485	488	564	5	1732
		48.9%	66.0%	68.2%	83.2%		68.5%
	DID NOT	204	250	228	114	3	796
		51.1%	34.0%	31.8%	16.8%		31.5%
	Missing	88	88	75	26	4	281
	TOTAL	399	735	716	678	12	2528
		100.0%	100.0%	100.0%	100.0%		

People with a college degree are more likely (83.2 percent) than those without a high school degree (48.9 percent) to report voting in 1996 (V = .24**).

Data File: **GSS**
Task: **Cross-tabulation**
Row Variable: **29) VOTE IN 96**
➤ Column Variable: **26) FAM INCOME**
➤ View: **Tables**
➤ Display: **Column %**

VOTE IN 96   by   FAM INCOME
Cramer's V: 0 173 **

		FAM INCOME				
		$0K-24.9K	$25K-49.9K	$50K +	Missing	TOTAL
VOTE IN 96	VOTED	463	437	628	209	1528
		59.7%	66.1%	78.6%		68.4%
	DID NOT	312	224	171	92	707
		40.3%	33.9%	21.4%		31.6%
	Missing	117	62	42	60	281
	TOTAL	775	661	799	361	2235
		100.0%	100.0%	100.0%		

Perhaps reflecting the education difference, the highest-income group is more likely than the lowest to report voting (V = .17**).

What difference, if any, did race and gender make?

	Data File:	**GSS**
	Task:	**Cross-tabulation**
	Row Variable:	**29) VOTE IN 96**
➤	Column Variable:	**20) RACE**
➤	View:	**Tables**
➤	Display:	**Column %**

VOTE IN 96   by   RACE
Cramer's V: 0.050 *

		RACE			
		BLACK	WHITE	Missing	TOTAL
VOTE IN 96	VOTED	247	1445	45	1692
		64.3%	70.7%		69.7%
	DID NOT	137	600	62	737
		35.7%	29.3%		30.3%
	Missing	48	193	40	281
	TOTAL	384	2045	147	2429
		100.0%	100.0%		

Race is only weakly related to voting (V = .05*) as whites were slightly more likely than African Americans to report voting in 1996.

	Data File:	**GSS**
	Task:	**Cross-tabulation**
	Row Variable:	**29) VOTE IN 96**
➤	Column Variable:	**19) GENDER**
➤	View:	**Tables**
➤	Display:	**Column %**

VOTE IN 96   by   GENDER
Cramer's V: 0.010

		GENDER		
		FEMALE	MALE	TOTAL
VOTE IN 96	VOTED	984	753	1737
		68.9%	68.0%	68.5%
	DID NOT	444	355	799
		31.1%	32.0%	31.5%
	Missing	160	121	281
	TOTAL	1428	1108	2536
		100.0%	100.0%	

No gender difference in voting appears (V = n.s.).

The voting literature also finds that most people who are politically alienated are less likely to vote. One measure of alienation is the belief that people cannot be trusted.

	Data File:	**GSS**
	Task:	**Cross-tabulation**
	Row Variable:	**29) VOTE IN 96**
➤	Column Variable:	**69) TRUSTED**
➤	View:	**Tables**
➤	Display:	**Column %**

VOTE IN 96   by   TRUSTED
Cramer's V: 0.152 **

		TRUSTED			
		CAN TRUST	BE CAREFUL	Missing	TOTAL
VOTE IN 96	VOTED	479	606	652	1085
		77.4%	62.9%		68.6%
	DID NOT	140	357	302	497
		22.6%	37.1%		31.4%
	Missing	43	130	108	281
	TOTAL	619	963	1062	1582
		100.0%	100.0%		

As expected, the alienated ("be careful") are less likely to report voting in 1996 (V = .15**).

If they did vote, were they especially likely to vote for Ross Perot, the major third-party candidate that year?

Data File: **GSS**

Task: **Cross-tabulation**

➤ Row Variable: **30) WHO IN 96?**

➤ Column Variable: **69) TRUSTED**

➤ View: **Tables**

➤ Display: **Column %**

WHO IN 96?   by   TRUSTED
Cramer's V: 0.106 **

		TRUSTED			
		CAN TRUST	BE CAREFUL	Missing	TOTAL
WHO IN 96?	CLINTON	247	307	364	554
		53.8%	54.4%		54.2%
	DOLE	167	166	165	333
		36.4%	29.4%		32.6%
	PEROT	45	91	80	136
		9.8%	16.1%		13.3%
	Missing	203	529	453	1185
	TOTAL	459	564	1062	1023
		100.0%	100.0%		

The alienated ("be careful") were slightly more likely than the nonalienated ("can trust") to vote for Perot.

What other factors affected candidate preferences in 1996? Let's try family income.

Data File: **GSS**

Task: **Cross-tabulation**

Row Variable: **30) WHO IN 96?**

➤ Column Variable: **26) FAM INCOME**

➤ View: **Tables**

➤ Display: **Column %**

WHO IN 96?   by   FAM INCOME
Cramer's V: 0.115 **

		FAM INCOME			
		$0K-24.9K	$25K-49.9K	$50K +	Missing	TOTAL
WHO IN 96?	CLINTON	294	230	293	101	817
		67.3%	55.4%	48.3%		56.0%
	DOLE	93	130	221	54	444
		21.3%	31.3%	36.5%		30.5%
	PEROT	50	55	92	19	197
		11.4%	13.3%	15.2%		13.5%
	Missing	455	308	235	187	1185
	TOTAL	437	415	606	361	1458
		100.0%	100.0%	100.0%		

All income groups were most likely to vote for Clinton; however, if you look within the table, you will see that most of the support for both Dole and Perot was from the middle and highest income groups. Conversely, Clinton received a majority of his support from the lowest income group.

What about race? Here we'll just compare whites and African Americans.

Data File: **GSS**

Task: **Cross-tabulation**

Row Variable: **30) WHO IN 96?**

➤ Column Variable: **20) RACE**

➤ View: **Tables**

➤ Display: **Column %**

WHO IN 96?   by   RACE
Cramer's V: 0.323 **

		RACE			
		BLACK	WHITE	Missing	TOTAL
WHO IN 96?	CLINTON	224	664	30	888
		94.1%	49.1%		55.8%
	DOLE	9	479	10	488
		3.8%	35.4%		30.7%
	PEROT	5	209	2	214
		2.1%	15.5%		13.5%
	Missing	194	886	105	1185
	TOTAL	238	1352	147	1590
		100.0%	100.0%		

African Americans overwhelmingly voted for Clinton (V = .32**). Why did such a large racial difference exist?

Have you heard of the gender gap in politics? Let's see how much of a gender gap there was in the 1996 election.

Data File: **GSS**
Task: **Cross-tabulation**
Row Variable: **30) WHO IN 96?**
➤ Column Variable: **19) GENDER**
➤ View: **Tables**
➤ Display: **Column %**

WHO IN 96?　by　GENDER
Cramer's V: 0.155 **

		GENDER		
		FEMALE	MALE	TOTAL
WHO IN 96?	CLINTON	574	344	918
		62.9%	47.8%	56.3%
	DOLE	227	271	498
		24.9%	37.7%	30.5%
	PEROT	112	104	216
		12.3%	14.5%	13.2%
	Missing	675	510	1185
	TOTAL	913	719	1632
		100.0%	100.0%	

Women were more likely (62.9 percent) than men (47.8 percent) to vote for Clinton (V = .16**). How would you explain this difference?

Let's turn to political ideology and see whether Americans have become more conservative since the early 1970s.

➤ Data File: **HISTORY**
➤ Task: **Historical Trends**
➤ Variable: **18) POL.VIEW**

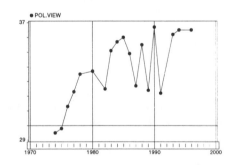

Percent saying their views are conservative

Americans are slightly more conservative than they were 30 years ago.

Several factors may influence how liberal or conservative people are. Let's look first at gender.

➤ Data File: **GSS**
➤ Task: **Cross-tabulation**
➤ Row Variable: **31) POL. VIEW**
➤ Column Variable: **19) GENDER**
➤ View: **Tables**
➤ Display: **Column %**

POL. VIEW　by　GENDER
Cramer's V: 0.083 **

		GENDER		
		FEMALE	MALE	TOTAL
POL. VIEW	LIBERAL	408	292	700
		27.7%	25.0%	26.5%
	MODERATE	621	433	1054
		42.1%	37.0%	39.9%
	CONSERV.	445	445	890
		30.2%	38.0%	33.7%
	Missing	114	59	173
	TOTAL	1474	1170	2644
		100.0%	100.0%	

Discovering Sociology

Men are slightly more conservative than women (V = .08**). Why do you think this is so?

Should race make a difference?

Data File: **GSS**
Task: **Cross-tabulation**
Row Variable: **31) POL. VIEW**
➤ Column Variable: **20) RACE**
➤ View: **Tables**
➤ Display: **Column %**

POL. VIEW   by   RACE
Cramer's V: 0.075 **

		RACE			
		BLACK	WHITE	Missing	TOTAL
POL. VIEW	LIBERAL	108	554	38	662
		27.3%	26.2%		26.4%
	MODERATE	183	810	61	993
		46.3%	38.3%		39.5%
	CONSERV.	104	753	33	857
		26.3%	35.6%		34.1%
	Missing	37	121	15	173
	TOTAL	395	2117	147	2512
		100.0%	100.0%		

Whites are somewhat more conservative than African Americans (V = .08**).

Now let's look at education. What difference, if any, do you think education will make?

Data File: **GSS**
Task: **Cross-tabulation**
Row Variable: **31) POL. VIEW**
➤ Column Variable: **15) EDUCATION**
➤ View: **Tables**
➤ Display: **Column %**

POL. VIEW   by   EDUCATION
Cramer's V: 0.120 **

		EDUCATION					
		NOT H.S.	H.S. GRAD	SOME COLL.	COLL. GRAD	Missing	TOTAL
POL. VIEW	LIBERAL	102	168	188	238	4	696
		24.2%	22.0%	24.7%	34.4%		26.4%
	MODERATE	195	368	292	195	4	1050
		46.3%	48.2%	38.4%	28.2%		39.8%
	CONSERV.	124	227	280	258	1	889
		29.5%	29.8%	36.8%	37.3%		33.7%
	Missing	66	60	31	13	3	173
	TOTAL	421	763	760	691	12	2635
		100.0%	100.0%	100.0%	100.0%		

The more educated the respondent, the more conservative, and also perhaps, the more liberal (V = .12**). Did you predict this result?

Finally, let's see whether religiosity is related to political ideology. What do you think we'll find?

Data File: **GSS**
Task: **Cross-tabulation**
Row Variable: **31) POL. VIEW**
➤ Column Variable: **54) PRAY**
➤ View: **Tables**
➤ Display: **Column %**

POL. VIEW   by   PRAY
Cramer's V: 0.137 **

		PRAY				
		DAILY	WEEKLY	< WEEKLY	Missing	TOTAL
POL. VIEW	LIBERAL	170	73	115	342	358
		23.4%	26.5%	39.0%		27.6%
	MODERATE	275	116	127	536	518
		37.8%	42.2%	43.1%		39.9%
	CONSERV.	283	86	53	468	422
		38.9%	31.3%	18.0%		32.5%
	Missing	50	11	26	86	173
	TOTAL	728	275	295	1432	1298
		100.0%	100.0%	100.0%		

People who pray daily are about twice as likely as those who pray less than weekly to be conservative (V = .14**). Why do you think this is so?

REVIEW QUESTIONS

Based on the first part of this exercise, answer True or False to the following items:

The most democratic nations tend to be in North America and Western Europe.	T	F
Political interest is highest in North America.	T	F
In the United States, the midwestern states were the most likely to vote for George Bush in 2000.	T	F
The poorest states were especially likely to vote for Al Gore in 2000.	T	F
In the GSS, African Americans were much more likely than whites to vote for Bill Clinton in 1996.	T	F
In the GSS, men were more likely than women to vote for Clinton in 1996.	T	F

EXPLORIT QUESTIONS

1. In the preliminary section of this exercise, we examined several variables in the GSS data file to see who was more likely to vote for Bill Clinton in 1996. Review the cross-tabulations in that discussion and then answer the following questions.

    a. In the preliminary section we examined several variables to see who was more likely to vote for Bill Clinton in 1996. Review the cross-tabulations in that discussion, and then indicate which pair of social background characteristics below was most likely to produce a vote for Clinton.
        1. Black, female
        2. Black, male
        3. White, female
        4. White, male

    b. Select one of the social backgrounds you just identified and briefly explain why you think people from that background were more likely to vote for Clinton.

2. Now determine whether religiosity was related to voting for Clinton in 1996. Obtain a cross-tabulation where 30) WHO IN 96? is the dependent (row) variable and 52) ATTEND is the independent (column) variable.

    a. What percent of people who never attend religious services voted for Clinton?   _____%

    b. What percent of people who attend religious services at least weekly voted for Clinton?   _____%

    c. Complete this sentence: The lower the religious attendance, the _____ likely people were to vote for Clinton.

3. The preliminary section also examined several variables to see what kinds of people were more likely to say that they are politically conservative.

    a. Review the cross-tabulations in that discussion and then indicate which pair of social background characteristics below was most likely to lead people to say they are conservative.
        1. Black, female
        2. Black, male
        3. White, female
        4. White, male

    b. Select one of the social backgrounds you just identified and briefly explain why you think people from that background are more likely to be conservative.

4. Now determine whether another measure of religiosity, religious attendance, is related to political conservatism. Obtain a cross-tabulation where 31) POL. VIEW is the dependent (row) variable and 52) ATTEND is the independent (column) variable.

    a. What percent of people who never attend religious services are conservative?   _____%

    b. What percent of people who attend religious services at least weekly are conservative?   _____%

    c. Complete this sentence: The lower the religious attendance, the _____ likely people are to be politically conservative.

5. Recall from the preliminary section that men are more likely than women to be conservative. This question has you examine the relationship between gender and political views while controlling for race.

> Data File: **GSS**
> Task: **Cross-tabulation**
> Row Variable: **31) POL. VIEW**
> ➤ Column Variable: **19) GENDER**
> ➤ Control Variable: **20) RACE**
> ➤ View: **Tables**
> ➤ Display: **Column %**

a. Examine the relationship between gender and political views just among whites. Which gender is more conservative (or less liberal)?   Men   Women

b. Now examine the relationship between gender and political views just among African Americans. Which gender is more conservative (or less liberal)?   Men   Women

c. Is the gender difference among whites statistically significant?   Yes   No

d. Is the gender difference among African Americans statistically significant?   Yes   No

6. The events of September 11, 2001 obviously ignited much discussion and concern about terrorist attacks on the United States. A year before these tragic events, the GSS asked respondents whether they thought terrorist threats by foreigners were more likely than 10 years earlier, less likely, or about as likely.

a. Obtain a univariate distribution for 161) FRTERROR. What percent said the threat of terrorist attacks by foreigners was more likely than 10 years earlier?   _____ %

b. Do at least 15 percent say that this threat was less likely than 10 years earlier?   Yes   No

c. Do you think the percent saying "more likely" would be higher today than in 2000, when the GSS asked this question? Why or why not?

7. Now determine whether education is related to concern about terrorist attacks as measured by the question you just examined by using 15) EDUCATION in a cross-tabulation.

a. What percent of college graduates say that a terrorist attack is more likely?   _____ %

b. What percent of high school dropouts say that a terrorist attack is more likely?   _____ %

c. Taking into account V, is education related to the view that terrorist attacks by foreigners are more likely when the GSS was conducted in 2000 than 10 years earlier?   Yes   No

8. The NATIONS data set includes a measure of the percent of people identifying themselves as being on the political left.

> ➤ *Data File:* **NATIONS**
> ➤ *Task:* **Mapping**
> ➤ *Variable 1:* **58) % LEFTISTS**
> ➤ *View:* **List: Rank**

   a. Which nation has the highest percent of people identifying themselves as being on the political left? _____

   b. Which nation has the lowest percent of people identifying themselves as being on the political left? _____

   c. What is the percent for the United States? _____%

9. Select a variable from the NATIONS set that you think will predict the percent of people who identify themselves as being on the political left. Obtain the appropriate scatterplot and write a brief essay in which you identify the variable you selected, explain why you thought it would be related to the percent on the political left, and summarize the results of your scatterplot.

10. We hear a lot of criticism these days about the federal government. Some of it is directed at the White House, and other criticism is directed at the Congress. The U.S. Supreme Court also comes in for its share of criticism. The GSS asks respondents how much confidence they have in each of these three branches of government. Complete each of the following analyses and keep notes on the results.

> ➤ *Data File:* **GSS**
> ➤ *Task:* **Univariate**
> ➤ *Primary Variable:* **71) FED.GOV'T?**
> ➤ *View:* **Pie**

> *Data File:* **GSS**
> *Task:* **Univariate**
> ➤ *Primary Variable:* **75) CONGRESS?**
> ➤ *View:* **Pie**

Data File: **GSS**
Task: **Univariate**
➤ Primary Variable: **74) SUP.COURT?**
➤ View: **Pie**

a. Generally speaking, in which branch of the government do people have
    the *most* confidence? (Circle one)?

    Executive branch

    Congress

    Supreme Court

b. In which branch of the government do people have the *least* confidence?
    (Circle one)?

    Executive branch

    Congress

    Supreme Court

11. Finally, let's see how confidence in the White House and Congress has changed in the past 25 years.
    We'll examine changes in the percent of the GSS reporting a "great deal of confidence" in these two
    branches of the government.

    ➤ Data File: **HISTORY**
    ➤ Task: **Historical Trends**
    ➤ Variable 1: **19) FED. GOVT.**
    ➤ Variable 2: **20) CONGRESS**

a. Generally speaking, would you say that the trends for the White House and
    Congress are similar to each other, or are they very different from each other?
    (Circle one.)

    Similar

    Very different

b. Compare the changes depicted in the graphs with the events listed below them. (Hint: Click on
    the small lines that represent the years.) Does it seem as though the changes are related to any
    or all of the events? Choosing one or two events, write a brief essay that supports your view.

12. The preliminary section of this exercise, briefly discussed support for George W. Bush in the presidential election of 2000. Recall that 70) %GWBUSH00 in the STATES data set maps the percent of voters in each state that voted for Bush. Select a variable from the STATES data set that you think will predict the percent of voters for Bush. Obtain the appropriate scatterplot and write a brief essay in which you identify the variable you selected, explain why you thought it would be related to the percent who voted for Bush, and summarize the results of your scatterplot.

HEALTH, ILLNESS, AND MEDICINE

Tasks: Mapping, Scatterplot, Univariate, Cross-tabulation, Historical Trends
Data Files: NATIONS, STATES, GSS, HISTORY

Medicine is yet another institution that is very much a part of our lives. Most of us were born in hospitals, and many of us will die in hospitals or other health-care facilities. In between, we spend a lot of time in the doctor's office when we or our loved ones are sick or injured. We go to the pharmacy to get prescription medications and to any number of places to get over-the-counter products such as aspirin and cough medicine. One way or the other, medicine affects all of us.

As this discussion should suggest, health and illness are far more than the medical matters that most health professionals consider them. Sociologists and many public health professionals recognize that health and illness are social and cultural matters as well. As such, they reflect many of the other factors we've been examining throughout this workbook: social class, gender, race/ethnicity, age, religiosity, and so on. In some cases these effects stem from the physiological differences between, say, women and men or`the elderly and the non-elderly. But in other cases they reflect the fault lines of a society where people with different socioeconomic rankings have different life outcomes.

This exercise explores some important aspects of health and medicine in today's world. We will look both at indicators of health and illness and at views on important health and medical issues. This focus will help reinforce the sociologist's emphasis on the social and cultural aspects of health and medicine.

A CROSS-CULTURAL LOOK AT HEALTH AND ILLNESS

When we examined stratification and the economy in previous exercises, we saw that the underdeveloped nations are far worse off than the developed nations on many indicators of health and illness. To reinforce this point, we map one such indicator, infant mortality, below. Notice the huge gap between North America and Western Europe and much of the rest of the world.

> *Data File:* **NATIONS**
> > *Task:* **Mapping**
> *Variable 1:* **10) INF. MORTL**
> > *View:* **Map**

INF. MORTL -- NUMBER OF INFANT DEATHS PER 1,000 BIRTHS (SAUS, 2000)

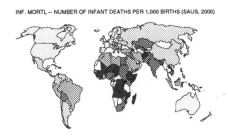

Ironically, however, some behaviors that put people at risk for illness and early death are more common in North America and Western Europe. Let's look at two of these risk factors.

Data File: **NATIONS**
Task: **Mapping**
➤ Variable 1: **108) CIGARETTES**
➤ View: **Map**

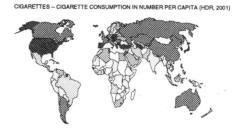

CIGARETTES -- CIGARETTE CONSUMPTION IN NUMBER PER CAPITA (HDR, 2001)

Cigarette smoking appears highest in North America, Europe, and parts of Asia. Despite their greater health problems in other respects, most of the underdeveloped nations have low rates of cigarette use.

Data File: **NATIONS**
Task: **Mapping**
Variable 1: **108) CIGARETTES**
➤ View: **List: Rank**

RANK	CASE NAME	VALUE
1	Greece	3923
2	Poland	3143
3	South Korea	2898
4	Japan	2857
5	Switzerland	2846
6	Singapore	2835
7	Bahrain	2819
8	Croatia	2632
9	Kuwait	2525
10	Czech Republic	2504

Greece has the dubious honor of leading the world in cigarette smoking.

Now let's look at alcohol use, which, if excessive, can lead to cirrhosis of the liver and other health problems.

Data File: **NATIONS**
Task: **Mapping**
➤ Variable 1: **104) ALCOHOL**
➤ View: **Map**

ALCOHOL -- NET ANNUAL ALCOHOL CONSUMPTION PER CAPITA, IN LITRES (IP)

Alcohol use appears most common in Europe and least common in the less developed nations elsewhere.

HEALTH AND MEDICINE IN THE UNITED STATES

At the state level we can see regional patterns of medical treatment and of illness. Let's start with AIDS, using the number of AIDS deaths per 100,000 population.

> ➤ *Data File:* **STATES**
> ➤ *Task:* **Mapping**
> ➤ *Variable 1:* **35) AIDS**
> ➤ *View:* **Map**

AIDS – 1998: AIDS/HIV DEATHS PER 100,000 (SA, 2000)

AIDS deaths are highest in the Northeast, a few southern states, and California, and lowest in the upper Midwest and the mountain states.

Although many people associate AIDS with homosexuality, AIDS also results from the sharing of needles in heroin use and from unprotected heterosexual sex. Do you think all three behaviors are more common in urban areas than in rural areas? If so, the more urban states should have higher AIDS death rates.

> *Data File:* **STATES**
> ➤ *Task:* **Scatterplot**
> ➤ *Dependent Variable:* **35) AIDS**
> ➤ *Independent Variable:* **19) %URBAN**
> ➤ *View:* **Reg. Line**

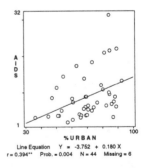

Line Equation   Y = -3.752 + 0.180 X
r = 0.394**   Prob. = 0.004   N = 44   Missing = 6

The more urban a state, the higher its AIDS death rate (r = .39**).

Being overweight is a health risk factor for many people. Which part of the country would you expect to be the heaviest?

Data File: **STATES**
➤ Task: **Mapping**
➤ Variable 1: **37) % FAT**
➤ View: **Map**

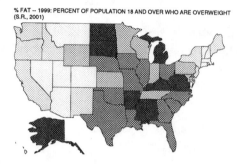

Obesity is more common east of the Mississippi. Why do you think this is so? Is it possible that more educated people are more aware of the health risk that obesity poses? If so, states with more college-educated people should have lower rates of obesity.

Data File: **STATES**
➤ Task: **Scatterplot**
➤ Dependent Variable: **37) % FAT**
➤ Independent Variable: **51) COLLEGE**
➤ View: **Reg. Line**

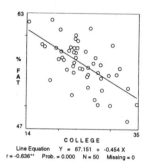

The scatterplot strongly supports the hypothesis (r = −.64**).

Should the more "overweight" states have higher rates of death by cardiovascular disease?

Data File: **STATES**
Task: **Scatterplot**
➤ Dependent Variable: **36) HEART DTHS**
➤ Independent Variable: **37) % FAT**
➤ View: **Reg. Line**

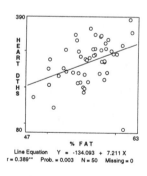

States with higher proportions of overweight people do have higher rates of cardiovascular deaths (r = .39**).

Many Americans lack health-care insurance. Let's see which parts of the country have the highest proportions of people without such insurance.

Data File: **STATES**
➤ Task: **Mapping**
➤ Variable 1: **41) HLTH INS**
➤ View: **Map**

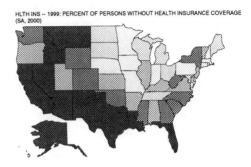

People in the South and the West are least likely to have health insurance. Should poorer states have greater proportions of people without medical insurance?

Data File: **STATES**
➤ Task: **Scatterplot**
➤ Dependent Variable: **41) HLTH INS**
➤ Independent Variable: **45) %POOR**
➤ View: **Reg. Line**

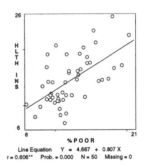

Most definitely (r = .61**).

We now continue our look at health and medicine with GSS data. We'll start with respondents' assessment of their own health.

➤ Data File: **GSS**
➤ Task: **Univariate**
➤ Primary Variable: **65) HEALTH**
➤ View: **Pie**

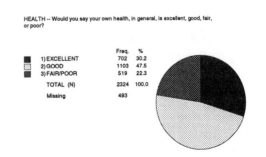

About 30 percent of the sample reports excellent health, and almost 48 percent reports good health. Twenty-two percent reports only fair or poor health.

In Exercise 6 we saw that poor people have worse health than the nonpoor. If that's true *only* because the poor are less educated, then if we control for education, income should no longer be related to health. That is, people with the same education should not differ in health even if they differ in income. Let's find out what happens when we do control for education.

Data File: **GSS**
➤ Task: **Cross-tabulation**
➤ Row Variable: **65) HEALTH**
➤ Column Variable: **26) FAM INCOME**
➤ Control Variable: **15) EDUCATION**
➤ View: **Tables (NOT H.S.)**
➤ Display: **Column %**

HEALTH by FAM INCOME
Controls: EDUCATION: NOT H.S.
Cramer's V: 0.178 **

		\$0K-24.9K	\$25K-49.9K	\$50K +	Missing	TOTAL
HEALTH	EXCELLENT	23	15	3	9	41
		9.7%	19.2%	13.0%		12.2%
	GOOD	93	39	17	29	149
		39.4%	50.0%	73.9%		44.2%
	FAIR/POOR	120	24	3	30	147
		50.8%	30.8%	13.0%		43.6%
	Missing	39	15	9	19	82
	TOTAL	236	78	23	87	337
		100.0%	100.0%	100.0%		

The option for selecting a control variable is located on the same screen you use to select other variables. For this example, select 15) EDUCATION as a control variable and then click [OK] to continue as usual. Separate tables for each of the 15) EDUCATION categories will now be shown for the 65) HEALTH and 26) FAM INCOME cross-tabulation. Examine these results before continuing.

➤ View: **Tables (H.S. GRAD)**
➤ Display: **Column %**

HEALTH by FAM INCOME
Controls: EDUCATION: H.S. GRAD
Cramer's V: 0.176 **

		\$0K-24.9K	\$25K-49.9K	\$50K +	Missing	TOTAL
HEALTH	EXCELLENT	43	54	54	20	151
		18.4%	26.9%	35.1%		25.6%
	GOOD	111	116	79	38	306
		47.4%	57.7%	51.3%		52.0%
	FAIR/POOR	80	31	21	26	132
		34.2%	15.4%	13.6%		22.4%
	Missing	49	39	41	21	150
	TOTAL	234	201	154	105	589
		100.0%	100.0%	100.0%		

Click the appropriate button at the bottom of the task bar to look at the second (or "next") partial table for 15) EDUCATION. Examine these results before continuing.

➤ View: **Tables (SOME COLL.)**
➤ Display: **Column %**

HEALTH by FAM INCOME
Controls: EDUCATION: SOME COLL.
Cramer's V: 0.188 **

		\$0K-24.9K	\$25K-49.9K	\$50K +	Missing	TOTAL
HEALTH	EXCELLENT	41	68	79	28	188
		21.4%	36.8%	40.5%		32.9%
	GOOD	91	95	95	37	281
		47.4%	51.4%	48.7%		49.1%
	FAIR/POOR	60	22	21	15	103
		31.3%	11.9%	10.8%		18.0%
	Missing	26	49	47	17	139
	TOTAL	192	185	195	97	572
		100.0%	100.0%	100.0%		

Again, click the appropriate button at the bottom of the task bar to look at the third (or "next") partial table for 15) EDUCATION. Examine these results before continuing.

➤ *View:* **Tables (COLL. GRAD)**
➤ *Display:* **Column %**

HEALTH   by   FAM INCOME
Controls: EDUCATION: COLL. GRAD
Cramer's V: 0.104 *

		\$0K-24.9K	\$25K-49.9K	\$50K +	Missing	TOTAL
HEALTH	EXCELLENT	31	65	139	29	235
		35.6%	47.1%	45.1%		44.1%
	GOOD	38	59	143	22	240
		43.7%	42.8%	46.4%		45.0%
	FAIR/POOR	18	14	26	4	58
		20.7%	10.1%	8.4%		10.9%
	Missing	26	17	61	12	116
	TOTAL	87	138	308	67	533
		100.0%	100.0%	100.0%		

FAM INCOME spans the columns \$0K-24.9K, \$25K-49.9K, \$50K +, Missing, TOTAL.

Click the appropriate button at the bottom of the task bar to look at the last (or "next") partial table for 15) EDUCATION. This table includes only college graduates.

Even when we control for education, income continues to be related to health: the lower the income, the greater the likelihood of only fair or poor health. While lower education may be one of the reasons for the poor health of poor people, other factors, such as lack of medical insurance or access to good medical care, must also be at work.

Some people think that the poor have worse health because they fail to seek medical treatment when they do have health problems. We can test this view indirectly with a GSS item that asks respondents how likely they would be to seek medical treatment if they had a health problem that prevented them from climbing several flights of stairs. If the poor have worse health because they do neglect to seek medical treatment, then we would expect to find that they would also be less likely to seek treatment in the hypothetical situation just posed.

Data File: **GSS**
Task: **Cross-tabulation**
➤ *Row Variable:* **156) TREAT2**
➤ *Column Variable:* **26) FAM INCOME**
➤ *View:* **Tables**
➤ *Display:* **Column %**

TREAT2   by   FAM INCOME
Cramer's V: 0.058

		\$0K-24.9K	\$25K-49.9K	\$50K +	Missing	TOTAL
TREAT2	LIKELY	275	255	257	110	787
		73.1%	78.9%	74.1%		75.2%
	NOT LIKELY	101	68	90	42	259
		26.9%	21.1%	25.9%		24.8%
	Missing	516	400	494	209	1619
	TOTAL	376	323	347	361	1046
		100.0%	100.0%	100.0%		

FAM INCOME spans the columns \$0K-24.9K, \$25K-49.9K, \$50K +, Missing, TOTAL.

There is *no* relationship between income and the willingness to seek medical treatment for the inability to climb several flights of stairs (V = .06).

Another GSS item asks how likely respondents would be to seek medical treatment if they had a health problem that limited their ability to accomplish as much as they would like with their work or other daily activities.

Data File: **GSS**
Task: **Cross-tabulation**
➤ *Row Variable:* **157) TREAT3**
➤ *Column Variable:* **26) FAM INCOME**
➤ *View:* **Tables**
➤ *Display:* **Column %**

TREAT3   by   FAM INCOME
Cramer's V: 0.004

		FAM INCOME				
		$0K-24.9K	$25K-49.9K	$50K +	Missing	TOTAL
TREAT3	LIKELY	272	228	247	101	747
		71.4%	71.3%	71.0%		71.2%
	NOT LIKELY	109	92	101	48	302
		28.6%	28.7%	29.0%		28.8%
	Missing	511	403	493	212	1619
	TOTAL	381	320	348	361	1049
		100.0%	100.0%	100.0%		

Once again there is no relationship between income and the willingness to seek medical treatment for the inability to accomplish as much as desired in work or other daily activities (V = .00 n.s.). These two cross-tabulations obviously do not disprove the belief that the poor neglect to seek medical treatment, but they do suggest that the belief may be mistaken.

As the United States continues to debate health-care policy, it's important to know public opinion on medicine and health care. The GSS asks how much confidence respondents have in medicine.

Data File: **GSS**
➤ *Task:* **Univariate**
➤ *Primary Variable:* **73) MEDICINE?**
➤ *View:* **Pie**

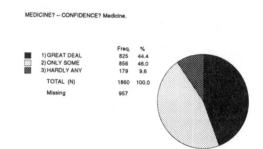

MEDICINE? -- CONFIDENCE? Medicine.

		Freq.	%
■	1) GREAT DEAL	825	44.4
▨	2) ONLY SOME	856	46.0
▨	3) HARDLY ANY	179	9.6
	TOTAL (N)	1860	100.0
	Missing	957	

About 44 percent of the public has a great deal of confidence in medicine, and 46 percent has some confidence. Almost 10 percent has hardly any confidence.

➤ *Data File:* **HISTORY**
➤ *Task:* **Historical Trends**
➤ *Variable:* **21) MEDICINE**

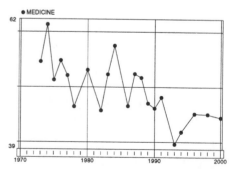

Percent having a great deal of confidence in medicine

The proportion expressing a great deal of confidence in medicine has declined a fair amount since the early 1970s. What might account for this decline?

WORKSHEET

NAME:

COURSE:

DATE:

EXERCISE

14

Workbook exercises and software are copyrighted. Copying is prohibited by law.

REVIEW QUESTIONS

Based on the first part of this exercise, answer True or False to the following items:

Alcohol use is highest in the most developed nations.   T   F

Cigarette use is generally highest in Africa.   T   F

In the United States, the reported AIDS rate is highest in the Midwest.   T   F

Despite expectations, the poorer states aren't more likely than richer states
to have people lacking medical insurance.   T   F

The higher the income the better the health.   T   F

In the GSS, confidence in medicine has risen overall since the 1970s.   T   F

EXPLORIT QUESTIONS

1. Should AIDS be more common in poorer nations?

> ➤ *Data File:* **NATIONS**
> ➤ *Task:* **Scatterplot**
> ➤ *Dependent Variable:* **101) AIDS**
> ➤ *Independent Variable:* **30) GDP/CAP**
> ➤ *View:* **Reg. Line**

a. Taking into account r, does this scatterplot support the hypothesis that AIDS is
more common in poorer nations?   Yes   No

b. Click on [Outlier]. What nation is the outlier that appears?   _____

c. Now remove this outlier. With this outlier removed, and taking into account r,
is the hypothesis now supported?   Yes   No

2. Use the [List:Rank] option to determine the names of the nations with the ten highest AIDS rates.

> *Data File:* **NATIONS**
> ➤ *Task:* **Mapping**
> ➤ *Variable 1:* **101) AIDS**
> ➤ *View:* **Map**

   a. In what continent are the majority of these nations located? (Note that you may
      use the [Find Case] option to locate these nations on the map.) (Circle one.)   North America

                                            South America

                                                 Europe

                                                    Africa

   b. Recalling our examination of this continent in Exercise 6 on social stratification,
      is this continent poor or relatively wealthy? (Circle one.)   Poor

                                                 Wealthy

   c. Thinking of your last two answers, is it accurate to say that the nations with the
      highest AIDS rates are poor nations?   Yes   No

   d. How does this conclusion help you to better understand the conclusion you drew in Question 1?

3. The preliminary section of this exercise, presented a scatterplot from the STATES data set of the relationship between obesity and cardiovascular deaths. Obtain this scatterplot once again and use the [Outlier] option to determine the state that is an outlier.

   a. Does the relationship become stronger or weaker if this outlier is removed?   Stronger   Weaker

   b. What state is the outlier?   _____

   c. Statistically speaking (i.e., in terms of its rankings on the two variables included in the scatter-plot), why is this state an outlier?

      1. It has a higher rate of cardiovascular deaths than would be expected from its level of obesity.

      2. It has a lower rate of cardiovascular deaths than would be expected from its level of obesity.

   d. Why do you think this particular state has the unusual sets of rankings that lead it to be an outlier?

4. The preliminary section also presented a map of the percent of each state's population who lack health insurance. What explains interstate variation in this percent? One possible factor is the percent of each state's population that is Latino, as the U.S. government estimates that about one-third of all Latinos lack health insurance. Perhaps the states with the highest Hispanic populations are also the states with the highest percent lacking health insurance.

> Data File: **STATES**
> Task: **Scatterplot**
> ➤ Dependent Variable: **41) HLTH INS**
> ➤ Independent Variable: **17) %HISPANIC**
> ➤ View: **Reg. Line**

a. What is the value of r? _____

b. Is r statistically significant? Yes   No

c. Do the results of this scatterplot support the hypothesis that states with higher Hispanic populations are also the states with the highest percent lacking health insurance? Yes   No

d. Locate New Mexico in the scatterplot. In which part of the scatterplot is it located? (Circle one.)

Upper left

Upper right

Lower left

Lower right

5. The Internet has become a major source of information for many topics and issues, including health. The GSS asked how often during the past month respondents have visited a Web site for information on health and fitness. Obtain a cross-tabulation where 176) WEB HEALTH is the dependent (row) variable and 19) GENDER is the independent (column) variable.

a. What percent of women have visited a Web site for health and fitness information at least three times in the past month? _____%

b. What percent of men have visited a Web site at least three times? _____%

c. Is V statistically significant? Yes   No

6. The preliminary section briefly discussed a GSS variable measuring the amount of confidence respondents place on the institution of medicine in the United States. Obtain a cross-tabulation where 73) MEDICINE? is the dependent (row) variable and 20) RACE the independent (column) variable.

a. What percent of blacks have a great deal of confidence in medicine? _____%

b. What percent of whites have a great deal of confidence in medicine? _____%

Exercise 14: Health, Illness, and Medicine

c. Is there a racial difference in how much confidence people have in medicine?   Yes   No

7. In the preliminary section of this exercise we mapped alcohol use. Let's see whether alcohol consumption at the nation level is related to the likelihood of cirrhosis of the liver. We'll first rank the deaths from this disease per 100,000 population and alcohol use.

> ➤ Data File: **NATIONS**
> ➤ Task: **Mapping**
> ➤ Variable 1: **97) CIRRHOSIS**
> ➤ Variable 2: **104) ALCOHOL**
> ➤ View: **Rank**

a. Do these ranked lists suggest these variables are fairly similar or very different? (Circle one.)

Fairly similar

Very different

b. Compare the two ranked lists. How many of the same nations appear in the top ten positions on both lists?

_____

8. Now let's obtain a scatterplot to reveal the association between alcohol use and cirrhosis deaths.

> Data File: **NATIONS**
> ➤ Task: **Scatterplot**
> ➤ Dependent Variable: **97) CIRRHOSIS**
> ➤ Independent Variable: **104) ALCOHOL**
> ➤ View: **Reg. Line**

Complete this sentence: The higher a nation's use of alcohol, the _____ its rate of cirrhosis.

9. In the average bar in the United States, it's probably true that many people are smoking cigarettes. When we look around the world, will we find that nations with high levels of alcohol consumption are also nations with high levels of cigarette use?

> Data File: **NATIONS**
> Task: **Scatterplot**
> ➤ Dependent Variable: **108) CIGARETTES**
> ➤ Independent Variable: **104) ALCOHOL**
> ➤ View: **Reg. Line**

a. Complete this sentence: The higher a nation's use of alcohol, the _____ its use of cigarettes.

    b. Does this scatterplot suggest that alcohol use leads to cigarette use?

        1. Yes, because alcohol use is related to cigarette use.

        2. No, because the use of both drugs might derive from another factor.

10. Syphilis is a very serious sexually transmitted disease. Which region of the United States will have the highest number of reported cases of syphilis per 100,000 population?

        ➤ *Data File:* **STATES**
        ➤ *Task:* **Mapping**
        ➤ *Variable 1:* **38) SYPHILIS**
        ➤ *View:* **Map**

    a. Which region has the highest rate? (Circle one.)

        Northeast

        South

        Midwest

        West

    b. Which region has the lowest rate? (Circle one.)

        Northeast

        South

        Midwest

        West

    c. What do you think accounts for the regional differences you just observed?

11. Locate a variable in the STATES data set that you think is related to a state's syphilis rate. Obtain a scatterplot where 38) SYPHILIS is the dependent variable and the variable you selected is the independent variable. Then answer the following questions.

    a. What independent variable did you select (name and number)? _____

    b. Why did you feel this variable would be associated with the syphilis rate?

c. What was the value of r in your scatterplot?   r = _____

d. Was r statistically significant? (Circle one.)   Yes   No

e. Was your hypothesis supported? (Circle one.)   Yes   No

◆ EXERCISE 15 ◆

POPULATION AND URBANIZATION

Tasks: Mapping, Scatterplot, Univariate, Cross-tabulation
Data Files: NATIONS, STATES, GSS

O ver the last several centuries, societies around the world have grown and become increasingly urbanized. People who formerly lived in small, isolated groups have moved closer together in greater and greater numbers. This process accelerated in the nineteenth century with the advent of industrialization. As people moved closer to the new economy's place of employment, the factory, cities grew by leaps and bounds. This growth had profound consequences for almost every aspect of social life.

Demography is the study of population growth and decline. Its key concepts are *fertility*, or the average number of children born to a woman; *mortality*, or the number of deaths in a society and usually measured as the death rate, or the number of deaths per 1,000 population; and *migration*, or the movement of people in and out of a specific society or location within that society. The net migration rate refers to the difference between the number of people moving in and the number moving out. Population growth or decline is a function of all three factors.

This exercise examines some key concepts and issues in the study of population and urbanization. Our primary focus will be on fertility, urbanization, and their correlates.

A DEMOGRAPHIC LOOK AROUND THE GLOBE

Demographers have long known that fertility is highest in the underdeveloped nations. Let's visualize the fertility differences they find.

> ➤ *Data File:* **NATIONS**
> ➤ *Task:* **Mapping**
> ➤ *Variable 1:* **8) FERTILITY**
> ➤ *View:* **Map**

FERTILITY – AVERAGE NUMBER OF CHILDREN BORN TO EACH WOMAN (SAUS, 2000)

This map provides one of the clearest distinctions between developed and underdeveloped nations that we've seen in this workbook. Fertility is generally highest in Africa and some Asian nations and lowest in North America and Europe.

Why do these fertility patterns exist? Demographers cite several reasons that we'll test separately. First, and in no particular order, women in underdeveloped nations are often prized above all for their

227

childbearing ability. Where women are so prized, fertility should be higher. Let's test both parts of this hypothesis before turning to the other reasons.

Data File: **NATIONS**
➤ Task: **Scatterplot**
➤ Dependent Variable: **52) HOME&KIDS**
➤ Independent Variable: **30) GDP/CAP**
➤ View: **Reg. Line**

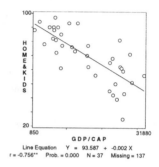

So far, so good. The less developed the nation, the more likely its citizens believe that a woman wants a home and children above all (r = −.76**). Let's now look at the relationship between the home/children variable and the actual fertility rate in nations.

Data File: **NATIONS**
Task: **Scatterplot**
➤ Dependent Variable: **8) FERTILITY**
➤ Independent Variable: **52) HOME&KIDS**
➤ View: **Reg. Line**

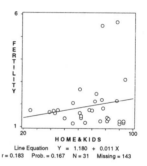

Not so good. The relationship between acceptance of women's traditional home role and fertility is in the expected direction but falls short of statistical significance (r = .18). One problem with using the independent variable chosen here is that it is from the World Values Study (WVS), which, although it includes most of the world's population in its 40-odd nations, doesn't include most of the underdeveloped nations, which typically have very small populations. Thus most WVS nations have relatively low fertility to begin with. Perhaps we'll have more success in illustrating the other reasons demographers cite for the high fertility of underdeveloped nations.

A second such reason is that contraception is uncommon in underdeveloped nations. When contraception is used less, fertility should obviously be higher.

Discovering Sociology

Data File: **NATIONS**
Task: **Scatterplot**
➤ Dependent Variable: **12) CONTRACEPT**
➤ Independent Variable: **30) GDP/CAP**
➤ View: **Reg. Line**

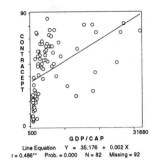

Line Equation   Y = 35.176 + 0.002 X
r = 0.486**   Prob. = 0.000   N = 82   Missing = 92

The less developed the nation, the lower its contraception use (r = .49**). Let's look at fertility rates and the use of contraception.

Data File: **NATIONS**
Task: **Scatterplot**
➤ Dependent Variable: **8) FERTILITY**
➤ Independent Variable: **12) CONTRACEPT**
➤ View: **Reg. Line**

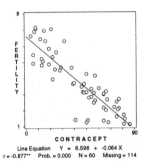

Line Equation   Y = 6.598 + -0.064 X
r = -0.877**   Prob. = 0.000   N = 60   Missing = 114

This is perhaps the strongest relationship we've seen in the workbook thus far (r = −.88**): the less the contraception use, the higher the fertility.

A third reason for high fertility in underdeveloped nations has to do with their national economies. Because the poorest nations are primarily agricultural, children are needed for economic reasons to help the family do its agricultural work. Let's see whether the agricultural nations have higher fertility.

Data File: **NATIONS**
Task: **Scatterplot**
Dependent Variable: **8) FERTILITY**
➤ Independent Variable: **34) % AGRIC $**
➤ View: **Reg. Line**

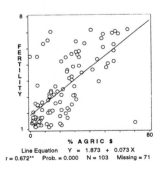

Line Equation   Y = 1.873 + 0.073 X
r = 0.672**   Prob. = 0.000   N = 103   Missing = 71

The agricultural nations are indeed much more likely to have higher fertility (r = .67**).

Exercise 15: Population and Urbanization

A final reason has to do with childhood mortality. In societies with high childhood mortality, families cannot assume that all their children will reach adulthood, and having greater numbers of children increases the chances of an adequate number's ultimately surviving. If this is true, high childhood mortality should be linked to higher fertility.

Data File: **NATIONS**
Task: **Scatterplot**
Dependent Variable: **8) FERTILITY**
➤ Independent Variable: **10) INF. MORTL**
➤ View: **Reg. Line**

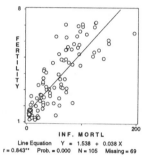

Nations with high childhood mortality are much more likely to have higher fertility (r = .85**).

None of the correlations just examined proves that the demographers' reasons for high fertility in underdeveloped nations are correct. In some of the correlations, causal order remains a question, and in some it's possible that the relationship is spurious if some third factor affects both the independent variable and fertility. But overall the correlations are consistent with the reasons demographers cite for the underdeveloped world–high fertility connection.

POPULATION AND URBANIZATION IN THE UNITED STATES

One of the most important population issues in the United States today is teenage pregnancies and births. To explore this issue, let's begin with the percent of births in each state to women under the age of 20.

➤ Data File: **STATES**
. ➤ Task: **Mapping**
➤ Variable 1: **39) TEEN MOM**
➤ View: **Map**

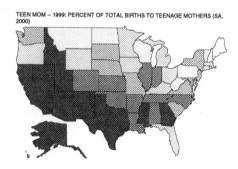

TEEN MOM -- 1999: PERCENT OF TOTAL BIRTHS TO TEENAGE MOTHERS (SA, 2000)

Teenage births are highest in the South and parts of the West.

What explains this geographic pattern? You've probably heard that most teenage mothers come from poor, uneducated families. Let's see whether the poorest and least-educated states have higher rates of teenage births.

Data File: **STATES**
➤ Task: **Scatterplot**
➤ Dependent Variable: **39) TEEN MOM**
➤ Independent Variable: **45) %POOR**
➤ View: **Reg. Line**

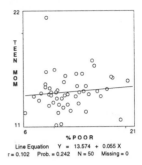

There are two outliers in this analysis, Utah and West Virginia. If you remove both of these outliers from the analysis you will find that the poorer the state, the higher its teenage birth rate (r = .35**). Let's see what the relationship is to education.

Data File: **STATES**
Task: **Scatterplot**
Dependent Variable: **39) TEEN MOM**
➤ Independent Variable: **52) HS DROPOUT**
➤ View: **Reg. Line**

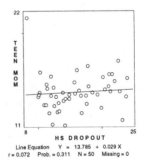

Once Utah and West Virginia are again removed as outliers, states with higher dropout rates tend to have higher teenage birth rates (.38**).

Time for the GSS. One GSS item asks respondents how many children they've ever had. Let's see what women report. (We'll use the [Subset] option to limit the analysis to women.)

> Data File: **GSS**
> Task: **Univariate**
> Primary Variable: **12) # CHILDREN**
> Subset Variable: **19) GENDER**
> Subset Category: **Include: 1) Female**
> View: **Pie**

CHILDREN -- How many children have you ever had? Please count all that were born alive at any time (including any you had from a previous marriage).

		Freq.	%
■	0) NONE	382	24.2
▨	1) 1-2	697	44.1
▨	2) 3 AND UP	502	31.8
	TOTAL (N)	1581	100.0
	Missing	7	

[Subset]

The option for selecting a subset variable is located on the same screen you use to select other variables. For this example, select 19) GENDER as the subset variable. A window will appear that shows you the categories of the subset variable. Select 1) Female as your subset category and choose the [Include] option. Then click [OK] and continue as usual.

Almost 32 percent of the women respondents have had three or more children, and another 44.1 percent have had 1 or 2 children. The remainder, 24.2 percent, have had no children.

What factors explain the number of children a woman has? We saw earlier that fertility was higher in poorer nations. Will GSS women whose parents had lower incomes be more likely than respondents with wealthier parents to have three or more children? Our independent (column) variable will be parents' occupational prestige.

> Data File: **GSS**
> Task: **Cross-tabulation**
> Row Variable: **12) # CHILDREN**
> Column Variable: **16) PARS.PRESG**
> Subset Variable: **19) GENDER**
> Subset Category: **Include: 1) Female**
> View: **Tables**
> Display: **Column %**

CHILDREN by PARS.PRESG

Cramer's V: 0.223 **

# CHILDREN	PARS.PRESG				
	BOTH LOW	BOTH MEDIU	BOTH HIGH	Missing	TOTAL
NONE	25	14	43	300	82
	18.7%	25.5%	45.3%		28.9%
1-2	56	22	40	579	118
	41.8%	40.0%	42.1%		41.5%
3 AND UP	53	19	12	418	84
	39.6%	34.5%	12.6%		29.6%
Missing	0	1	0	6	7
TOTAL	134	55	95	1303	284
	100.0%	100.0%	100.0%		

Note that you again need to select GENDER as a subset variable.

Respondents whose parents both held low-prestige jobs (and presumably had lower incomes) are three times more likely than those whose parents both held high-prestige jobs to have three or more children (V = .22**).

Now let's consider the woman's own schooling and hypothesize that women with less education will have more children.

Data File: **GSS**
Task: **Cross-tabulation**
Row Variable: **12) # CHILDREN**
➤ Column Variable: **15) EDUCATION**
➤ Subset Variable: **19) GENDER**
➤ Subset Category: **Include: 1) Female**
➤ View: **Tables**
➤ Display: **Column %**

CHILDREN   by   EDUCATION

Cramer's V: 0.179 **

		NOT H.S.	H.S. GRAD	SOME COLL.	COLL. GRAD	Missing	TOTAL
# CHILDREN	NONE	39	92	110	141	0	382
		14.3%	18.7%	24.3%	39.6%		24.3%
	1-2	104	226	214	149	4	693
		38.2%	45.9%	47.3%	41.9%		44.1%
	3 AND UP	129	174	128	66	5	497
		47.4%	35.4%	28.3%	18.5%		31.6%
	Missing	2	3	2	0	0	7
	TOTAL	272	492	452	356	9	1572
		100.0%	100.0%	100.0%	100.0%		

Women with a college degree are much less likely than those without a high school degree to have three or more children, and substantially more likely to have no kids (V = .18**). Notice that causal order here is unclear, because it's possible that having children restricted a woman's ability to go to, and graduate from, college. Still, most demographers feel that education does help explain how many children a woman has.

What about religious preference? Because the Catholic Church disapproves of birth control, do you think Catholic women have more children than Protestant or Jewish women?

Data File: **GSS**
Task: **Cross-tabulation**
Row Variable: **12) # CHILDREN**
➤ Column Variable: **51) RELIGION**
➤ Subset Variable: **19) GENDER**
➤ Subset Category: **Include: 1) Female**
➤ View: **Tables**
➤ Display: **Column %**

CHILDREN   by   RELIGION

Cramer's V: 0.074 **

		PROTESTANT	CATHOLIC	JEWISH	Missing	TOTAL
# CHILDREN	NONE	174	104	11	93	289
		19.2%	26.5%	39.3%		21.8%
	1-2	416	166	12	103	594
		46.0%	42.3%	42.9%		44.9%
	3 AND UP	314	122	5	61	441
		34.7%	31.1%	17.9%		33.3%
	Missing	5	1	0	1	7
	TOTAL	904	392	28	258	1324
		100.0%	100.0%	100.0%		

Catholics do *not* have more children than Protestants. In fact, they tend to have fewer children than Protestants (V = .07**).

Now let's turn our attention to people living in cities and in rural areas. We'll look at medium- and large-city residents who have lived in the same city all their lives and compare them to rural residents who have lived in the same rural town or farm all their lives.

Going back to at least Ferdinand Toennies a century ago, sociologists have depicted cities as cold, alienating places with weak social bonds and high amounts of distrust. Other sociologists have criticized this view as an unfair stereotype and point to closely knit neighborhoods within large cities. Let's see which view, if either, the GSS data support.

The GSS asks respondents how often they "spend a social evening" with a neighbor. If the critics of cities are right, urban respondents should spend less time than rural respondents with a neighbor.

Data File: **GSS**
Task: **Cross-tabulation**
➤ Row Variable: **78) SOC.NEIGH.**
➤ Column Variable: **141) URBAN?**
➤ View: **Tables**
➤ Display: **Column %**

SOC.NEIGH. by URBAN?
Cramer's V: 0.145 *

		URBAN?			
		URBAN	RURAL	Missing	TOTAL
SOC.NEIGH.	DAILY/WKLY	70	41	505	111
		35.7%	41.4%		37.6%
	MONTH/YEAR	57	37	578	94
		29.1%	37.4%		31.9%
	NEVER	69	21	483	90
		35.2%	21.2%		30.5%
	Missing	106	54	796	956
	TOTAL	196	99	2362	295
		100.0%	100.0%		

Urban residents spend less time than rural residents with their neighbors (V = .14**). This table supports the city critics.

Another variable asks how often respondents spend a social evening with relatives.

Data File: **GSS**
Task: **Cross-tabulation**
➤ Row Variable: **77) SOC.KIN**
➤ Column Variable: **141) URBAN?**
➤ View: **Tables**
➤ Display: **Column %**

SOC.KIN by URBAN?
Cramer's V: 0.049

		URBAN?			
		URBAN	RURAL	Missing	TOTAL
SOC.KIN	DAILY/WKLY	131	69	841	200
		65.8%	69.7%		67.1%
	MONTH/YEAR	56	26	649	82
		28.1%	26.3%		27.5%
	NEVER	12	4	76	16
		6.0%	4.0%		5.4%
	Missing	103	54	796	953
	TOTAL	199	99	2362	298
		100.0%	100.0%		

Taking into account statistical significance, urban and rural residents are equally likely to spend time with relatives (V = .05). So, are the critics of cities correct or not?

Critics of cities would predict that urban residents are less trusting of others than rural residents are. Let's check this out by using responses to the question "Generally speaking, would you say that most people can be trusted or that you can't be too careful in dealing with people?"

Data File: **GSS**
Task: **Cross-tabulation**
➤ Row Variable: **69) TRUSTED**
➤ Column Variable: **141) URBAN?**
➤ View: **Tables**
➤ Display: **Column %**

TRUSTED by URBAN?
Cramer's V: 0.068

		URBAN?			
		URBAN	RURAL	Missing	TOTAL
TRUSTED	CAN TRUST	62	26	574	88
		32.3%	25.7%		30.0%
	BE CAREFUL	130	75	888	205
		67.7%	74.3%		70.0%
	Missing	110	52	900	1062
	TOTAL	192	101	2362	293
		100.0%	100.0%		

Again, taking into account statistical significance, urban and rural residents are equally distrustful (V = .07).

The GSS also asks, "Do you think most people would try to take advantage of you if they got a chance, or would they try to be fair?"

Data File: **GSS**
Task: **Cross-tabulation**
➤ Row Variable: **68) ADVANTAGE?**
➤ Column Variable: **141) URBAN?**
➤ View: **Tables**
➤ Display: **Column %**

ADVANTAGE?   by   URBAN?
Cramer's V: 0.025

		URBAN	RURAL	Missing	TOTAL
ADVANTAGE?	TAKE ADVAN	103	51	564	154
		56.3%	53.7%		55.4%
	BE FAIR	80	44	849	124
		43.7%	46.3%		44.6%
	Missing	119	58	949	1126
	TOTAL	183	95	2362	278
		100.0%	100.0%		

Both sets of residents are also equally likely to think that people take advantage (V = .02).

None of these last three tables supports the bleak view of the critics of cities.

WORKSHEET

NAME:

COURSE:

DATE:

EXERCISE

15

REVIEW QUESTIONS

Based on the first part of this exercise, answer True or False to the following items:

The more agricultural nations have higher fertility.	T	F
Fertility tends to be very high in Africa.	T	F
In the United States, teenage births are highest in the South and parts of the West.	T	F
In the United States, the poorer a state, the higher its percent of births to teenage mothers.	T	F
In the GSS, Catholic women have had more children than Protestant women.	T	F
GSS data support the common view that cities are cold, alienating places.	T	F

EXPLORIT QUESTIONS

1. At the international level, is fertility related to believing that the ideal family size is three or more children?

> ➤ Data File: **NATIONS**
> ➤ Task: **Scatterplot**
> ➤ Dependent Variable: **8) FERTILITY**
> ➤ Independent Variable: **9) LARGE FAML**
> ➤ Display: **Reg. Line**

a. Consider this summary of the scatterplot: Nations whose people are more likely to think that the ideal number of children is three or more also tend to be nations with _____ levels of fertility. Which word, higher or lower, should be inserted into the blank?　　　　　　　　　　　　　　　　　　　　　　Higher　　　Lower

b. This scatterplot suggests which of the following?
   1. A belief in large families affects fertility.
   2. Fertility affects beliefs about family size.
   3. The scatterplot suggests both of these causal possibilities.

2. If urban areas are indeed more cold and alienating, then urban states in the United States should have higher suicide rates.

> ➤ *Data File:* **STATES**
> ➤ *Task:* **Scatterplot**
> ➤ *Dependent Variable:* **80) SUICIDE**
> ➤ *Independent Variable:* **19) %URBAN**
> ➤ *Display:* **Reg. Line**

a. Is r statistically significant?   Yes   No

b. Are suicide rates higher in the more urbanized states?   Yes   No

c. Locate Nevada in the scatterplot. In which part of the scatterplot is Nevada located? (Circle one.)

     Upper left

     Upper right

     Lower left

     Lower right

3. An important variable in population studies is the age at which a woman has her first child. What factors might account for this age?

> ➤ *Data File:* **GSS**
> ➤ *Task:* **Cross-tabulation**
> ➤ *Row Variable:* **181) AGE KD BRN**
> ➤ *Column Variable:* **16) PARS.PRESG**
> ➤ *Subset Variable:* **19) GENDER**
> ➤ *Subset Category:* **Include: 1) Female**
> ➤ *View:* **Tables**
> ➤ *Display:* **Column %**

The option for selecting a subset variable is located on the same screen you use to select other variables. For this example, select 19) GENDER as the subset variable. A window will appear that shows you the categories of the subset variable. Select 1) Female as your subset category and choose the [Include] option. Then click [OK] and continue as usual.

a. What percent of female respondents with parents with high-prestige jobs had their first child before age 20?   _____%

b. What percent of female respondents with parents with low-prestige jobs had their first child before age 20?   _____%

c. Is V statistically significant?   Yes   No

d. Is parents' social class related to the age at which women have their
first child?   Yes   No

e. More than one-tenth of the women whose parents were in low-prestige jobs
had their first child after age 29.   T   F

4. In the preliminary section of this exercise, we saw that a woman's education was related to the number of children she has had. It makes to sense to think that her education should also be related to
the age at which she had her first child. Our hypothesis is that the lower her education, the more likely she had a child before she was 20-years-old.

> Data File: **GSS**
> Task: **Cross-tabulation**
> Row Variable: **181) AGE KD BORN**
> ➤ Column Variable: **15) EDUCATION**
> ➤ Subset Variable: **19) GENDER**
> ➤ Subset Category: **Include: 1) Female**
> ➤ View: **Tables**
> ➤ Display: **Column %**

a. What percent of women without a high school degree had their first
child before age 20?   _____%

b. What percent of women with a college degree had their first child
before age 20?   _____%

c. Is V statistically significant?   Yes   No

d. Is education related to the age at which women have their first child?   Yes   No

e. Discuss how and why this relationship is difficult to interpret in terms of causal order.

5. It is often thought that urban residents are more tolerant of controversial behaviors than rural residents. Let's test this assumption.

> Data File: **GSS**
> Task: **Cross-tabulation**
> ➤ Row Variable: **97) HOMO.SEX**
> ➤ Column Variable: **141) URBAN?**
> ➤ View: **Tables**
> ➤ Display: **Column %**

Exercise 15: Population and Urbanization   239

Notice, if you are continuing from the previous question, you must exit the task, delete the subset variable, or clear all variables to remove the subset variable.

a. What percent of urban residents think homosexuality is always wrong? _____%

b. What percent of rural residents think homosexuality is always wrong? _____%

c. Is V statistically significant? Yes   No

d. Are urban residents more tolerant of homosexuality? Yes   No

6. Now let's examine views on premarital sex.

> Data File: **GSS**
> Task: **Cross-tabulation**
> ➤ Row Variable: **94) PREM.SEX**
> ➤ Column Variable: **141) URBAN?**
> ➤ View: **Tables**
> ➤ Display: **Column %**

a. What percent of urban residents think premarital sex is always wrong? _____%

b. What percent of rural residents think premarital sex is always wrong? _____%

c. Is V statistically significant? Yes   No

d. Are urban residents more tolerant of premarital sex? Yes   No

7. Based on your answers to Questions 5 and 6, would you say that urban residents are more tolerant than rural residents of controversial behaviors? What differences between urban and rural areas might explain your answer?

8. Some scholars believe that to lower fertility we must first give women more power and freedom to control their lives. In this way of thinking, when women have options outside the home and can decide when they want to marry and have children, they will decide to have fewer babies and, con-comitantly, not be forced to conceive babies they don't want. If this is true, nations where women are the most empowered in these and other ways should have lower fertility. Let's test this hypothesis.

➤ *Data File:* **NATIONS**
➤ *Task:* **Scatterplot**
➤ *Dependent Variable:* **8) FERTILITY**
➤ *Independent Variable:* **48) FEM POWER**
➤ *Display:* **Reg. Line**

a. Taking into account r and its statistical significance, do the nations where women
   are most empowered have lower fertility? (Circle one.)   Yes   No

b. Locate the United States on the scatterplot and compare its position to the regression line. Does
   the United States have considerably more fertility than would be expected from its level of female
   empowerment, considerably less fertility, or about the expected fertility?

   1. Considerably more fertility

   2. Considerably less fertility

   3. About the expected fertility

9. In Questions 5–7 we examined whether urban residents in the GSS were more tolerant than rural
   residents of certain controversial behaviors. Let's explore the relationship between urbanism and tolerance with two behaviors included in the NATIONS data set.

   Data File: **NATIONS**
   Task: **Scatterplot**
   ➤ *Dependent Variable:* **96) PROSTITUTE**
   ➤ *Independent Variable:* **4) URBAN %**
   ➤ *Display:* **Reg. Line**

   Data File: **NATIONS**
   Task: **Scatterplot**
   ➤ *Dependent Variable:* **95) GAY SEX**
   ➤ *Independent Variable:* **4) URBAN %**
   ➤ *Display:* **Reg. Line**

   Do these scatterplots suggest that the more urban nations are more tolerant of
   controversial behaviors than the less urban nations?   Yes   No

10. Which part of the United States should have the highest birth rate?

    ➤ *Data File:* **STATES**
    ➤ *Task:* **Mapping**
    ➤ *Variable 1:* **27) BIRTHS**
    ➤ *View:* **Map**

    a. Generally speaking, which region has the highest birth rate?   _____

b. Why do you think this region has the highest rate?

c. Select a variable from the STATES data set that you think might be related to the states' birth rates. Obtain a scatterplot where 27) BIRTHS is the dependent variable and the variable you select is the independent variable. Then write a brief essay where you indicate which variable you selected, the reasons why you thought it would be related to the states' birth rates, and the results of your analysis.

11. Do religious women have more children than less religious women?

> Data File: **GSS**
> Task: **Cross-tabulation**
> Row Variable: **12) # CHILDREN**
> Column Variable: **54) PRAY**
> Subset Variable: **19) GENDER**
> Subset Category: **Include: 1) Female**
> View: **Tables**
> Display: **Column %**

a. What percent of women who pray daily have three or more children?   _____%

b. What percent of women who pray less than weekly have three or more children?   _____%

c. This analysis assumes that religiosity affects the number of children a woman has. Is it possible instead that the number of children affects religiosity? Explain your answer.

d. It is also possible that this relationship is spurious—that some third factor affects both religiosity and the number of children a woman has. What would be one possible such factor? Explain your answer.

SOCIAL CHANGE AND MODERNIZATION

Tasks: Scatterplot, Mapping, Historical Trends, Univariate, Cross-tabulation
Data Files: NATIONS, STATES, HISTORY, GSS

Much of this workbook has been about social change. Time after time we've seen how the wealthy, developed, industrial societies differ from the poor, underdeveloped, agricultural ones. In doing so, we've seen both the good and bad effects of modernization. In looking at the United States, we've seen that some attitudes and behaviors have changed in the last 30 years—admittedly a short time span—but that others have not. Many scholars trace the changes that have occurred to the turbulent 1960s, when the civil rights movement in the South, the Vietnam antiwar movement, and other social-change efforts altered the course of our nation.

This exercise summarizes some of the major changes industrialization has brought around the world and reviews some of the major ways in which the United States has changed since the 1960s. The worksheet questions address several different aspects of social change and modernization.

SOCIAL CHANGE AND MODERNIZATION
IN GLOBAL PERSPECTIVE

Modernization, as we've seen, has had important consequences for the life chances, behaviors, and attitudes of societies around the world. Let's review some of the changes, both good and bad, that modernization has brought. Our independent variable will be a nation's gross domestic product per capita. We'll use this variable to see what difference wealth and modernization makes in four broad areas of belief and practice.

Modernization and Life Chances

> *Data File:* **NATIONS**
> *Task:* **Scatterplot**
> *Dependent Variable:* **116) EDUCATION**
> *Independent Variable:* **30) GDP/CAP**
> *View:* **Reg. Line**

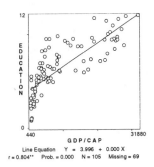

Line Equation   Y = 3.996 + 0.000 X
r = 0.804**   Prob. = 0.000   N = 105   Missing = 69

Wealthy societies have much higher levels of education (r = .80**).

Data File: **NATIONS**
Task: **Scatterplot**
➤ Dependent Variable: **10) INF. MORTL**
➤ Independent Variable: **30) GDP/CAP**
➤ View: **Reg. Line**

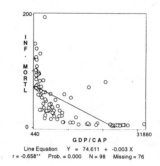

Wealthy societies also have much lower rates of infant mortality (r = −.66**) . . .

Data File: **NATIONS**
Task: **Scatterplot**
➤ Dependent Variable: **11) MOM MORTAL**
➤ Independent Variable: **30) GDP/CAP**
➤ View: **Reg. Line**

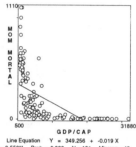

. . . much lower rates of death during childbirth (r = −.55**) . . .

Data File: **NATIONS**
Task: **Scatterplot**
➤ Dependent Variable: **17) DEATH RATE**
➤ Independent Variable: **30) GDP/CAP**
➤ View: **Reg. Line**

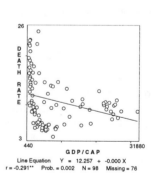

. . . and lower rates of death from all causes (r = −.29**).

But wealth and modernization also have their personal costs.

Data File: **NATIONS**
Task: **Scatterplot**
➤ Dependent Variable: **104) ALCOHOL**
➤ Independent Variable: **30) GDP/CAP**
➤ View: **Reg. Line**

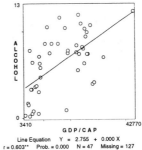

People in the wealthiest nations are apt to drink more alcohol (r = .60**) . . .

Data File: **NATIONS**
Task: **Scatterplot**
➤ Dependent Variable: **102) DRUGS**
➤ Independent Variable: **30) GDP/CAP**
➤ View: **Reg. Line**

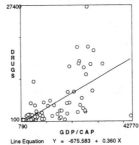

. . . to use narcotic drugs (r = .62**) . . .

Data File: **NATIONS**
Task: **Scatterplot**
➤ Dependent Variable: **26) MEAT CONS.**
➤ Independent Variable: **30) GDP/CAP**
➤ View: **Reg. Line**

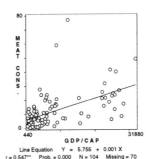

. . . to eat (high-fat!) meat (r = .55**) . . .

Exercise 16: Social Change and Modernization

Data File: **NATIONS**	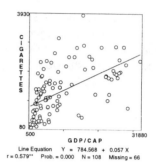
Task: **Scatterplot**	
➤ *Dependent Variable:* **108) CIGARETTES**	
➤ *Independent Variable:* **30) GDP/CAP**	
➤ *View:* **Reg. Line**	

... and to smoke cigarettes (r = .58**).

Modernization, Technology, and the Environment

Not surprisingly, wealthy nations are especially likely to enjoy the benefits of modern technology, but those benefits, too, have a cost, in this case a global one.

Data File: **NATIONS**	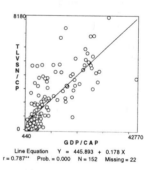
Task: **Scatterplot**	
➤ *Dependent Variable:* **44) TLVSN/CP**	
➤ *Independent Variable:* **30) GDP/CAP**	
➤ *View:* **Reg. Line**	

On the upside, wealthy societies are much more likely to have such things as televisions (r = .79**) . . .

Data File: **NATIONS**	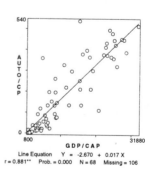
Task: **Scatterplot**	
➤ *Dependent Variable:* **40) AUTO/CP**	
➤ *Independent Variable:* **30) GDP/CAP**	
➤ *View:* **Reg. Line**	

. . . and cars (r = .88**).

248   *Discovering Sociology*

But all of that technology has an environmental cost.

Data File: **NATIONS**

Task: **Scatterplot**

➤ Dependent Variable: **33) ELECTRIC**

➤ Independent Variable: **30) GDP/CAP**

➤ View: **Reg. Line**

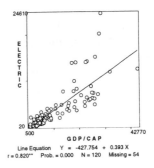

Line Equation   Y = -427.754 + 0.393 X
r = 0.820**   Prob. = 0.000   N = 120   Missing = 54

Wealthy societies use much more electricity than traditional ones (r = .82**) . . .

Data File: **NATIONS**

Task: **Scatterplot**

➤ Dependent Variable: **119) GREENHOUSE**

➤ Independent Variable: **30) GDP/CAP**

➤ View: **Reg. Line**

➤ Find: **Outlier/Remove**

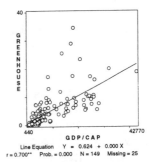

Line Equation   Y = 0.624 + 0.000 X
r = 0.700**   Prob. = 0.000   N = 149   Missing = 25

Find the outlier by selecting the [Outlier] option. A box will appear around the dot representing the outlier case. Remove this case by clicking the [Remove] button. Notice the change in the r value after removing the outlier case.

. . . and contribute more to global emissions (r = .70**) and thus to a possible greenhouse effect.

Modernization and Gender, Racial, and Ethnic Inequality

Modernization brings with it not only better life chances overall and improved technology, but also greater gender equality and less racial and ethnic prejudice.

Data File: **NATIONS**

Task: **Scatterplot**

➤ Dependent Variable: **48) FEM POWER**

➤ Independent Variable: **30) GDP/CAP**

➤ View: **Reg. Line**

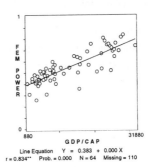

Line Equation   Y = 0.383 + 0.000 X
r = 0.834**   Prob. = 0.000   N = 64   Missing = 110

Wealthy societies have much more gender equality than poorer societies (r = .83**).

Data File: **NATIONS**

Task: **Scatterplot**

➤ Dependent Variable: **52) HOME&KIDS**

➤ Independent Variable: **30) GDP/CAP**

➤ View: **Reg. Line**

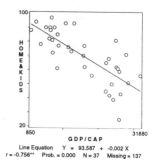

In line with this fact, people in the wealthier nations are much less likely to feel a woman's place is in the home (r = −.76**).

The NATIONS data set doesn't have measures of racial equality, but does include measures of racial and ethnic prejudice.

Data File: **NATIONS**

Task: **Scatterplot**

➤ Dependent Variable: **80) RACISM**

➤ Independent Variable: **30) GDP/CAP**

➤ View: **Reg. Line**

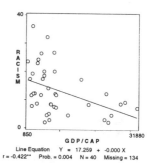

Looking at one such measure, people in the wealthier nations are much less likely to say they wouldn't want members of other races as their neighbors (r = −.42**). Racial prejudice declines as societies become more modern and wealthy.

Modernization and Tolerance for Controversial Behaviors

The last area in which we'll examine modernization's effects is tolerance. One hallmark of modern societies is greater tolerance of some of the behaviors that less modern societies often condemn. Although disapproval of such behaviors may still be common in modern societies, it's less common than in the less modern ones.

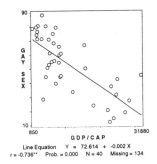

Data File: **NATIONS**
Task: **Scatterplot**
➤ Dependent Variable: **95) GAY SEX**
➤ Independent Variable: **30) GDP/CAP**
➤ View: **Reg. Line**

People in the wealthier nations are less apt to think that homosexuality is never acceptable (r = −.74**).

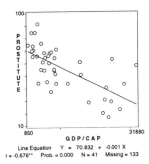

Data File: **NATIONS**
Task: **Scatterplot**
➤ Dependent Variable: **96) PROSTITUTE**
➤ Independent Variable: **30) GDP/CAP**
➤ View: **Reg. Line**

They're also less likely to think that prostitution is never acceptable (r = −.68**).

SOCIAL CHANGE IN THE UNITED STATES

We turn now to the GSS to explore a few ways in which Americans have changed their attitudes and behaviors in the last quarter century, and a few areas in which they've not changed. This period of our nation's history followed on the heels of the tumultuous 1960s and thus represents a fascinating time in which to explore whether the '60s movements and other events have had an enduring impact. The last 30 years have also seen significant political developments and controversies that may have changed people's views or practices.

Sex and Drugs

We'll begin our exploration with a look at views and practices having to do with sex and drugs.

➤ *Data File:* **HISTORY**
   ➤ *Task:* **Historical Trends**
➤ *Variable:* **17) HOMO.SEX**

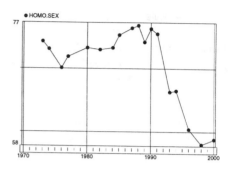

Percent saying homosexual sex is always wrong

The percent saying homosexuality is "always wrong" stayed fairly stable from the early 1970s through the early 1990s, but has declined since that time.

Data File: **HISTORY**
   Task: **Historical Trends**
➤ *Variable:* **16) PREM.SEX**

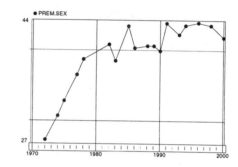

Percent saying premarital sex is not wrong

The percent saying premarital sex is "not wrong" rose through the 1970s and early 1980s and remains higher today than 30 years ago.

Data File: **HISTORY**
   Task: **Historical Trends**
➤ *Variable:* **15) XMAR.SEX**

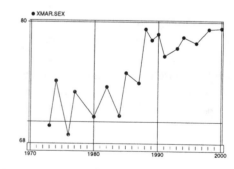

Percent saying extramarital sex is always wrong

Although views on homosexuality and premarital sex have become more lenient in the past 30 years, views on extramarital sex—adultery—have become more intolerant.

Data File: **HISTORY**
Task: **Historical Trends**
➤ Variable: **25) ABORT ANY**

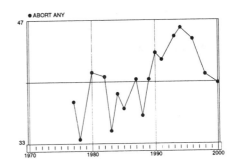

Percent favoring legal abortion for any reason

Support for legalized abortion for any reason rose from the late 1980s through the mid-1990s but has declined since.

Data File: **HISTORY**
Task: **Historical Trends**
➤ Variable: **4) GRASS?**

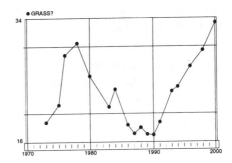

Percent saying marijuana should be made legal

The percent thinking marijuana should be made legal rose rapidly from the early 1970s through the late 1970s and then declined, only to rise again during the 1990s.

Data File: **HISTORY**
Task: **Historical Trends**
➤ Variable: **26) SMOKE?**

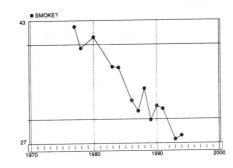

Percent smoking cigarettes

Cigarette smoking dropped considerably from the mid-1970s through the mid-1990s.

Data File: **HISTORY**
Task: **Historical Trends**
➤ Variable: **27) DRINK?**

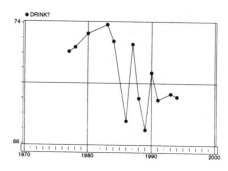

Percent drinking alcohol instead of abstaining

Drinking, however, has declined only slightly.

Political Beliefs

The second area we examine concerns political beliefs. We'll start with the percent who agree that "most public officials are not really interested in the problems of the average person."

Data File: **HISTORY**
Task: **Historical Trends**
➤ Variable: **28) ANOMIA 7**

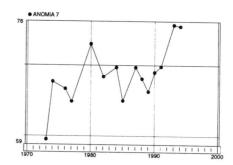

Percent thinking politicians don't care about what the average person thinks

This percent has fluctuated since the early 1970s. It rose in the latter half of that decade and declined during Ronald Reagan's presidency, then rose again. It remained higher in 1994, when the GSS last asked this question, than at the beginning of the 1970s.

Has the United States become more conservative during the past 30 years? To find out, let's see whether today's GSS respondents view themselves as more conservative than their counterparts 30 years ago.

Data File: **HISTORY**
Task: **Historical Trends**
➤ Variable: **18) POL.VIEW**

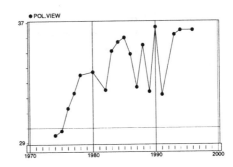

Percent saying their views are conservative

The country has become slightly more conservative in the past 30 years.

Gender and Racial Issues

Our third area is gender and racial issues.

Data File: **HISTORY**
Task: **Historical Trends**
➤ Variable: **31) WOMAN PRES**

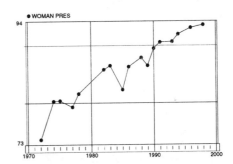

Percent willing to vote for a qualified woman for president

The percent of the public saying it would vote for a qualified woman for president has risen substantially since the early 1970s.

Data File: **HISTORY**
Task: **Historical Trends**
➤ Variable: **34) WOMEN WORK**

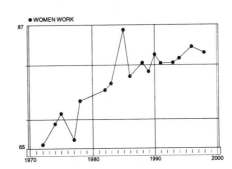

Percent approving of a married woman working outside of the home

So has the percent approving of a married woman working outside the home.

Data File: **HISTORY**
Task: **Historical Trends**
➤ Variable: **7) BLACK PRES**

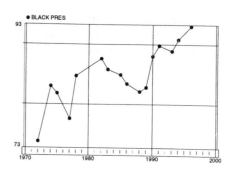

Percent willing to vote for a qualified African American for president

Despite some fluctuations, the percent of the public saying it would vote for a qualified African American for president rose substantially after the early 1970s.

Data File: **HISTORY**
Task: **Historical Trends**
➤ Variable: **8) RACE SEG**

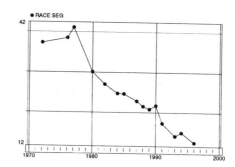

Percent favoring racial segregation in housing

Meanwhile, agreement that white people have a right to keep African Americans out of their neighborhoods dropped substantially.

Data File: **HISTORY**
Task: **Historical Trends**
➤ Variable: **6) INTERMAR.?**

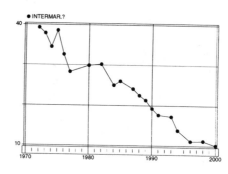

Percent agreeing with laws against racial intermarriage

As has the percent supporting laws against racial intermarriage.

Discovering Sociology

Religion and Religious Issues

Our last area concerns religion.

Data File: **HISTORY**
Task: **Historical Trends**
➤ Variable: **32) ATTEND**

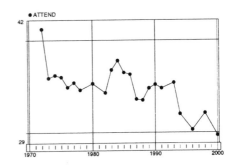

Percent attending religious services at least once per week

Weekly religious attendance has generally declined since the early 1970s.

Data File: **HISTORY**
Task: **Historical Trends**
➤ Variable: **33) SCH.PRAYER**

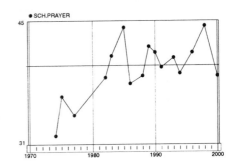

Percent approving Supreme Court decision banning prayer in public schools

Support for the U.S. Supreme Court's prohibition of prayer in the public schools has risen slightly during this period.

Data File: **HISTORY**
Task: **Historical Trends**
➤ Variable: **36) ATHEIST SP**

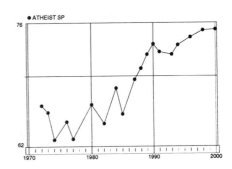

Percent approving of the right of an atheist to make a speech

Support for the right of an atheist to make a speech has also risen.

NAME:

COURSE:

DATE:

REVIEW QUESTIONS

Based on the first part of this exercise, answer True or False to the following items:

The more modern a nation, the more likely its residents are to eat meat.	T	F
The more modern a nation, the greater its gender equality.	T	F
During the 1980s, U.S. disapproval of homosexuality dropped substantially.	T	F
Judging from GSS data in the HISTORY file, Americans have become more racially prejudiced since the 1970s.	T	F
Judging from GSS data in the HISTORY file, Americans have become less sexist since the 1970s.	T	F
Religious attendance has increased since the early 1970s.	T	F

EXPLORIT QUESTIONS

1. The brief exploration earlier of changes in several kinds of beliefs and practices during the past 30 years indicates that Americans have, overall, slightly abandoned traditional beliefs and practices on matters such as sexuality and drugs, gender, race, and religion. Do you think the changes we've seen in this exercise have been good or bad for the United States? Why do you feel the way you do?

2. The preliminary section indicated that one cost of modernization is damage to the environment. The quality of the environment is one of the most important issues in the United States today and around the world. Many Americans are members of groups concerned about the environment or have otherwise been active in environmental protection efforts. The GSS includes a measure of whether respondents have done at least one of the following: (a) belonged to or given money to an environmental group or (b) signed a petition or taken part in a protest about an environmental issue. To determine whether age predicts involvement in such environmental activism, obtain a cross-tabulation where 139) ENV ACTIVE is the dependent (row) variable and 14) AGE 65+ is the independent (column) variable.

Exercise 16: Social Change and Modernization

a. What percent of people under age 65 have engaged in environmental activism as measured by this question?   _____%

b. What percent of people age 65 and older have engaged in environmental activism?   _____%

c. Complete this sentence: People 65 and older are _____ percent less likely than those under 65 to have engaged in environmental activism.

3. As the previous question indicated, scholars of political participation and social movements study the factors that predict greater or less involvement in political activism. One factor they emphasize is education. Generally speaking, the higher the education, the greater the involvement in many kinds of political activities. Our hypothesis will therefore be that the higher the education, the greater the likelihood of being involved in the type of environmental activism measured by 139) ENV ACTIVE. To test this hypothesis, obtain a cross-tabulation where 139) ENV ACTIVE is the dependent variable and 15) EDUCATION is the independent variable.

   a. What percent of college graduates have engaged in environmental activism as measured by this question?   _____%

   b. What percent of high school dropouts have engaged in environmental activism?   _____%

   c. Complete this sentence: College graduates are _____ percent more likely than high school dropouts to have engaged in environmental activism.

4. Scholars also emphasize that people will be more likely to be activists if they are especially concerned about a particular issue. The GSS asks respondents whether they agree or disagree that "many of the claims about environmental threats are exaggerated." Respondents who disagreed with this statement are evidently more concerned about the environment than respondents who agreed with this statement. Our hypotheses will therefore be that respondents who disagreed with the statement will be more likely to engage in environmental activities than those who agreed with it.

   a. Obtain a cross-tabulation where 139) ENV ACTIVE is the dependent (row) variable and 158) GREEN EXAG is the independent (column) variable.

   b. What percent of respondents who disagree that environmental threats are exaggerated have engaged in environmental activism?   _____%

   c. What percent of respondents who agree that these threats are exaggerated have engaged in environmental activism?   _____%

   d. Do these results support the view that people are more likely to be activists if they are especially concerned about a particular issue?   Yes   No

5. Select another variable from the GSS that you think will be related to whether people have engaged in environmental activism. Obtain a cross-tabulation where 139) ENV ACTIVE is the dependent (row) variable and the variable you select is the independent (column) variable.

    a. What variable (number and name) did you select?   _____

    b. Why did you think this variable would be related to participation in environmental activism?

    c. Briefly summarize the results of your cross-tabulation and draw a conclusion regarding whether these results support your hypothesis.

6. The GSS also asks respondents whether they think that the United States is doing "more than enough, about right, or too little" to protect the world environment. We will hypothesize that liberals are more likely than conservatives to think that the United States is doing too little to protect the world environment. To test this hypothesis, obtain a cross-tabulation where 177) US ENVIRO is the dependent (row) variable and 31) POL. VIEW is the independent (column) variable.

    a. What percent of liberals think the U.S. is doing too little to protect the world environment?   _____%

    b. What percent of conservatives think the U.S. is doing too little to protect the world environment?   _____%

    c. Taking into account V, do these results support the hypothesis?   Yes   No

7. The NATIONS data set includes a measure of the percent who feel "the entire way our society is organized must be radically changed by revolutionary action." Scholars of revolution often theorize that this view is more common in societies with worse living conditions. To test this hypothesis, obtain a scatterplot where 57) REVOLUTION is the dependent variable and 24) QUAL. LIFE is the independent variable.

➤ *Data File:*   **NATIONS**
➤ *Task:*   **Scatterplot**
➤ *Dependent Variable:*   **57) REVOLUTION**
➤ *Independent Variable:*   **24) QUAL. LIFE**
➤ *Display:*   **Reg. Line**

a. What is the value of r?   r = _____

b. Which statement below best summarizes the results of this scatterplot?

    1.   The better the quality of life, the more likely people are to believe revolution is necessary.

    2.   The worse the quality of life, the more likely people are to believe revolution is necessary.

    3.   The worse the quality of life, the less likely people are to believe revolution is necessary.

c. Since the United States has very good living conditions compared to much of the rest of the world, we would expect Americans to disagree with the statement that revolution is necessary. If so, where should we expect the United States to appear in the scatterplot? (Circle one.)

                                                             Upper left

                                                           Upper right

                                                             Lower left

                                                           Lower right

d. Now locate the United States in the scatterplot. In which part of the scatterplot does it appear? (Circle one.)

                                                             Upper left

                                                             Upper right

                                                             Lower left

                                                             Lower right

e. Does the location of the United States support the theory that revolutionary views are more common in societies with worse living conditions?   Yes   No

8. If, as we saw in the preliminary section, wealthy nations drink more, are their citizens also more likely to die from cirrhosis of the liver?

         Data File: **NATIONS**
         ➤ *Task:*   **Scatterplot**
         ➤ *Dependent Variable:*   **97) CIRRHOSIS**
         ➤ *Independent Variable:*   **30) GDP/CAP**
         ➤ *Display:*   **Reg. Line**

a. What is the value of r?   r = _____

b. Is r statistically significant?   Yes   No

c. Locate the United States on the scatterplot. Judging from its position relative to the regression line, does the United States have a considerably higher rate of cirrhosis than would be expected from its wealth, a considerably lower rate, or about the expected rate?

1. Considerably higher

2. Considerably lower

3. About the expected rate

9. In the preliminary section we also saw that the more modern nations were less racially prejudiced than the less modern nations. Are they also less anti-Semitic?

> Data File: **NATIONS**
> Task: **Scatterplot**
> ➤ Dependent Variable: **77) ANTI-SEM.**
> ➤ Independent Variable: **30) GDP/CAP**
> ➤ Display: **Reg. Line**

In the space below, summarize what the results of this scatterplot indicate.

10. In the United States, important population shifts occurred during the 1990s.

> ➤ Data File: **STATES**
> ➤ Task: **Mapping**
> ➤ Variable 1: **5) POP GROW**
> ➤ View: **Map**

a. What part of the country generally experienced the greatest population growth during the 1990s?   _____

b. What part of the country generally experienced the least population growth during the 1990s?   _____

c. How would you explain these results?

11. Some states gained in population much more than others.

> *Data File:* **STATES**
> *Task:* **Mapping**
> *Variable 1:* **5) POP GROW**
> ➤ *View:* **List: RANK**

    a. Which state had the highest population growth?   _____

    b. Which state had the least population growth?   _____

    c. What ranking did your home state achieve?   _____

12. In the preliminary section we saw that attitudes on sexual matters have generally become more tolerant in the United States in the last 30 years. What about views on sex education? Use the HISTORY data set and HISTORICAL TRENDS task to obtain a graph of changes during this period in the percent approving of sex education in the schools. (Hint: When selecting variables, use the [Search] feature to locate the appropriate item.)

    a. Is approval for sex education now higher than in the early 1970s, lower, or about the same? (Circle one.)

                                                              Higher

                                                              Lower

                                                      About the same

    b. From the late 1970s to the late 1980s, did approval for sex education rise, fall, or stay about the same? (Circle one.)

                                                               Rise

                                                               Fall

                                                     Stay about the same

    c. What might account for the changes depicted in the graph?

APPENDIX: VARIABLE NAMES AND SOURCES

Note for MicroCase Users: These data files may be used with the MicroCase Analysis System. If you are moving variables from these files into other MicroCase files, or vice versa, you may need to reorder the cases. Also note that files that have been modified in MicroCase will not function in Student ExplorIt.

◆ DATA FILE: GSS ◆

1) FULL TIME?
2) PART TIME?
3) RETIRED
4) UNEMPLOYED
5) HOURS
6) MARITAL
7) DIVORCED?
8) MATE WORK?
9) DAD OCCUP
10) MA OCC. 80
11) # SIBS
12) # CHILDREN
13) AGE
14) AGE 65+
15) EDUCATION
16) PARS.PRESG
17) SOUTH
18) REGION
19) GENDER
20) RACE
21) PARS. DIV?
22) ALONE?
23) HOMEMAKER?
24) MA WRK GRW
25) BORN USA?
26) FAM INCOME
27) OWN INCOME
28) POL. PARTY
29) VOTE IN 96
30) WHO IN 96?
31) POL. VIEW
32) ENVIRON. $
33) HEALTH $
34) BIG CITY $
35) CRIME $
36) EDUCATE $
37) BLACK $
38) DEFENSE $
39) FOR. AID $

40) WELFARE $2
41) ATHEIST SP
42) RACIST SP
43) COMMUN. SP
44) MILIT SP
45) HOMO. SP
46) HOMO. TCH
47) EXECUTE?
48) GUN LAW?
49) COURTS?
50) GRASS?
51) RELIGION
52) ATTEND
53) HOW RELIG?
54) PRAY
55) RELIG. 1ST
56) SCH. PRAY
57) BIBLE
58) INTERMAR.?
59) RACE PUSH
60) BL.IN AREA
61) AFFRM. ACT
62) BLKS IMP
63) HAPPY?
64) HAP.MARR.?
65) HEALTH
66) LIFE
67) HELPFUL?
68) ADVANTAGE?
69) TRUSTED
70) BIG BIZ?
71) FED.GOV'T?
72) LABOR?
73) MEDICINE?
74) SUP. COURT?
75) CONGRESS?
76) MILITARY?
77) SOC.KIN
78) SOC.NEIGH.

79) SOC.FRIEND
80) SOC. BAR
81) LIVE W KID
82) LOSE JOB?
83) WRK IF $$
84) SELF RANK?
85) SAT.$?
86) UNIONIZED?
87) EVER UNEMP
88) GET AHEAD?
89) MEN BETTER
90) ABORT ANY
91) IDEAL#KIDS
92) TEEN BC OK
93) DIV.EASY?
94) PREM.SEX
95) TEEN SEX
96) XMAR.SEX
97) HOMO.SEX
98) SPANKING
99) EUTHANASIA
100) SUIC.ILL
101) OWN GUN?
102) WATCH TV
103) MOTH.WORK?
104) PRESCH.WRK
105) WIFE@HOME
106) RACE DIF1
107) RACE DIF2
108) RACE DIF3
109) RACE DIF4
110) GOV.BLACK
111) WHITE WORK
112) JEW WORK
113) BLACK WORK
114) ASIAN WORK
115) HISP. WORK
116) LIVE JEW
117) LIVE BLACK

◆ DATA FILE: GSS (cont'd) ◆

118) LIVE ASIAN
119) LIVE HISP
120) MARRY JEW
121) MARRY BLK
122) MARRY ASN
123) MARRY HSP
124) REV.DISCRM
125) MELTNG.POT
126) IMMIGRANTS
127) DISC MAN
128) DISC WOMAN
129) HIRE WOMEN
130) PARTNER
131) TOGETHER
132) GOOD LIFE
133) INEQUAL 3
134) INEQUAL 4
135) GOVT EQUAL
136) MEN.OVRWRK
137) ENV.HARMS
138) ENV.PROGRS
139) ENV ACTIVE

140) RECYCLE
141) URBAN?
142) SX.PRTNRS?
143) SEX FREQ.
144) EVER STRAY
145) CONDOM
146) #CRCT. WORD
147) MUSLM CONT
148) IMM NO JOB
149) LETIN HISP
150) TIMEKID
151) SAT DEMOC?
152) LEFT ALONE
153) DOWN BLUE
154) SOC ACTS
155) TREAT1
156) TREAT2
157) TREAT3
158) GREEN EXAG
159) EVER CRACK
160) USTERROR
161) FRTERROR

162) CHANGE $?
163) PARS.DEGR.
164) DENOM
165) CHOICE
166) GREEN GRP?
167) GREEN MONY
168) FEAR WALK
169) KNW WHITE
170) KNW BLACK
171) KNW JEW
172) KNW HISP
173) KNW ASIAN
174) LIKE JOB?
175) PRESTIGE
176) WEB HEALTH
177) US ENVIRO
178) RACEGENDER
179) INEQUAL 5
180) SEX ED?
181) AGE KD BRN

◆ DATA FILE: NATIONS ◆

1) COUNTRY
2) POPULATION
3) DENSITY
4) URBAN %
5) POP GROWTH
6) NETMIGRT
7) BIRTH RATE
8) FERTILITY
9) LARGE FAML
10) INF. MORTL
11) MOM MORTAL
12) CONTRACEPT
13) ABORTION
14) ABORT LEGL
15) MOM HEALTH
16) AB. UNWANT
17) DEATH RATE
18) LIFE EXPCT
19) SEX RATIO
20) % UNDER 15
21) % OVER 65

22) HUMAN DEV.
23) ECON DEVEL
24) QUAL. LIFE
25) CALORIES
26) MEAT CONS.
27) %UNDRWGHT
28) DOCTORS
29) INEQUALITY
30) GDP/CAP
31) UNEMPLYRT
32) ENG C/CP
33) ELECTRIC
34) % AGRIC $
35) % INDUS $
36) % SERVC $
37) %WORK AG
38) %WORK IN
39) $WORK OT
40) AUTO/CP
41) TELPH/CP
42) LITERACY

43) RADIO/CP
44) TLVSN/CP
45) CELL PHONE
46) %FEM.LEGIS
47) M/F EDUC.
48) FEM POWER
49) SEX MUTIL
50) SINGLE MOM
51) WORK MOM?
52) HOME&KIDS
53) WED PASSE'
54) FREEDOM
55) %TURNOUT
56) MULTI-CULT
57) REVOLUTION
58) % LEFTISTS
59) %RIGHTISTS
60) P.INTEREST
61) PETITION?
62) BOYCOTT?
63) DEMONSTRAT

Discovering Sociology

◆ DATA FILE: NATIONS (cont'd) ◆

64) DEFNS/CP
65) %MUSLIM
66) %CHRISTIAN
67) %CATHOLIC
68) %HINDU
69) %BUDDHIST
70) %JEWISH
71) REL.PERSON
72) CH.ATTEND
73) GOD IMPORT
74) PRAY?
75) PRISONERS
76) CAP PUNISH
77) ANTI-SEM.
78) ANTI-FORGN
79) ANTI-MUSLM
80) RACISM
81) ANTI-GAY
82) HOME LIFE?

83) CHORES?
84) WORK PRIDE
85) WORK IMPT?
86) UNIONIZED?
87) UNIONS?
88) POOR LAZY
89) INJUSTICE
90) TRUST?
91) CHEAT TAX
92) TAKE BRIBE
93) EX-MARITAL
94) MINOR SEX
95) GAY SEX
96) PROSTITUTE
97) CIRRHOSIS
98) SUICIDE
99) SUICIDE NO
100) EUTHANASIA
101) AIDS

102) DRUGS
103) SMOKE DOPE
104) ALCOHOL
105) SPIRITS
106) BEER DRINK
107) WINE DRINK
108) CIGARETTES
109) DO SPORTS?
110) VERY HAPPY
111) FAMILY IMP
112) KID MANNER
113) KID INDEPN
114) KID OBEY
115) KID THRIFT
116) EDUCATION
117) LIFEX MALE
118) LIFEX FEMALE
119) GREENHOUSE

◆ DATA FILE: STATES ◆

1) STATE NAME
2) SOUTHNESS
3) WESTNESS
4) WARM WINTR
5) POP GROW
6) POPULATION
7) NDM 90–99
8) %UND 18
9) %OVER 64
10) DENSITY
11) %MALE
12) %FEMALE
13) %WHITE
14) %BLACK
15) %AMER.IND.
16) %ASIAN/PI
17) %HISPANIC
18) NOT DENSE
19) %URBAN
20) %RURAL
21) %IRISH
22) %ITALIAN
23) IMMIGRANTS
24) MARRIAGES

25) DIVORCES
26) F HEAD/C
27) BIRTHS
28) SMALL B/W
29) NO PHONE
30) %NORELIG
31) % JEWISH
32) % CATHOLIC
33) % BAPTIST
34) INF. MORT.
35) AIDS
36) HEART DTHS
37) % FAT
38) SYPHILIS
39) TEEN MOM
40) ABORTIONS
41) HLTH INS
42) HMO
43) DOCTORS
44) PICKUPS
45) %POOR
46) MED.FAM $
47) PER CAP$
48) POOR>65

49) UNEMP
50) %UNION
51) COLLEGE
52) HS DROPOUT
53) BOYS' LIFE
54) V.CRIME
55) P.CRIME
56) MURDER
57) RAPE
58) ROBBERY
59) ASSAULT
60) BURGLARY
61) LARCENY
62) CAR THEFTS
63) PRISON
64) DEATH PENL
65) ROLL.STONE
66) COSMO
67) PLAYBOY
68) F&STREAM
69) STATES 00
70) %GWBUSH00
71) %GORE00
72) %NADER00

◆ DATA FILE: STATES (cont'd) ◆

73) STATES '96
74) %CLINTON96
75) %DOLE 96
76) %PEROT 96
77) %VOTED 96

78) %VOTED 00
79) VETERANS
80) SUICIDE
81) MOBILITY
82) %NURS.HOME

83) CH.MEMBERS
84) %TV>6HRS
85) %AGRI.EMP
86) PUB.TRANS.
87) AVG. TRAVL

◆ DATA FILE: HISTORY ◆

1) Date
2) UNIONIZED
3) LABOR?
4) GRASS?
5) FEAR WALK
6) INTERMAR.?
7) BLACK PRES
8) RACE SEG
9) RACE DIF2
10) RACE DIF4
11) WOMEN HOME
12) EUTHANASIA
13) LIVE W KID

14) HAP.MARR
15) XMAR.SEX
16) PREM.SEX
17) HOMO.SEX
18) POL.VIEW
19) FED. GOVT.
20) CONGRESS
21) MEDICINE
22) GOV. MED
23) # CHILDREN
24) SEX ED.
25) ABORT ANY
26) SMOKE?

27) DRINK?
28) ANOMIA 7
29) ANOMIA 6
30) POL. PARTY
31) WOMAN PRES
32) ATTEND
33) SCH.PRAYER
34) WOMEN WORK
35) WIFE@HOME
36) ATHEIST SP
37) YEAR

SOURCES

GSS

The GSS data file is based on selected variables from the National Opinion Research Center (University of Chicago) General Social Survey for 2000, distributed by the Roper Center and the Inter-University Consortium for Political and Social Research. The GSS is sponsored by the National Science Foundation, and the principal investigators are James A. Davis and Tom W. Smith.

HISTORY

The data in the HISTORY file come from all the administrations of the General Social Survey since its inception in 1972 through the latest available administration in 2000.

NATIONS

The data in the NATIONS file are from a variety of sources. The variable description for each variable uses the following abbreviations to indicate the source.

FITW: Freedom in the World, published annually by Freedom House

HDR: Human Development Report, published annually by the United Nations Development Program

IDB: International Data Base, 1998, U.S. Bureau of the Census

IDEA: Institute for Democracy and Electoral Assistance. Turnout data are from the institute's "Global Report on Political Participation." (Stockholm, 1997) Electorial system data and coding from the "International Handbook of Electoral System Design" (Stockholm, 1997).

IP: International Profile: Alcohol and Other Drugs, published by the Alcoholism and Drug Addiction Research Foundation (Toronto), 1994

KIDRON & SEGAL: State of the World Atlas, 5th Edition, London: Penguin, 1995

NBWR: The New Book of World Rankings, 3d Edition, Facts on File, 1991

PON: The Progress of Nations, UNICEF, 1996

SAUS: Statistical Abstract of the United States, published annually by the U.S. Department of Commerce

STARK: Coded and calculated by Rodney Stark

TWF: The World Factbook, published annually by the Central Intelligence Agency

TWW: The World's Women, published by the United Nations, 1995

UNCRIME: United Nations. THE FOURTH UNITED NATIONS SURVEY OF CRIME TRENDS AND OPERATIONS OF CRIMINAL JUSTICE SYSTEMS, 1986–1990 (computer files). Vienna, Austria: Crime Prevention and Criminal Justice Branch, United Ntions Office at Vienna, 1994. The United States did not provide data for the Fourth Survey. Therefore, the crime rates for the United States were taken from the Statistical Abstract of the United States, 1996.

UNSY: United Nations Statistical Yearbook, 1997, United Nations

WCE: World Christian Encyclopedia, David R. Barrett, editor, Oxford University Press, 1982

WDI: World Development Indicators, published annually by the World Bank

WVS: World Values Study Group. WORLD VALUES SURVEY, 1981–1984, 1990–1993, and 1995–1997 (computer files). ICPSR version. Ann Arbor, MI: Institute for Social Research (producer), 1998. Ann Arbor, MI: Inter-University Consortium for Political and Social Research (distributor), 1998.

STATES

The data in the STATES file are from a variety of sources. The variable description for each variable uses the following abbreviations to indicate the source.

ABC: Audit Bureau of Circulation Blue Book for the indicated year

BJS: Bureau of Justice Statistics

CENSUS: The summary volumes of the U.S. Census for the indicated year

CHURCH: Churches and Church Membership in the United States, published every 10 years by the Glenmary Research Center, Atlanta, for the year indicated

FEC: Federal Election Commission for the year indicated

HIGHWAY: Federal Highway Administration, Highway Statistics for the indicated year

KOSMIN: Kosmin, Barry A. and Egon Mayer. 2001 American Religious Identification Survey, New York: CUNY Graduate Center

NCHS: National Center for Health Statistics

SA: Statistical Abstract of the United States for the indicated year

SMAD: State and Metropolitan Area Data Book for the indicated year

S.R.: State Rankings (Morgan Quitno Corp., Lawrence, KS) for the indicated year

UCR Uniform Crime Reports for the indicated year